Vibrationally Mediated Photodissociation

Vibrationally Mediated Photodissociation

Salman (Zamik) Rosenwaks
Department of Physics, Ben-Gurion University of the Negev, Beer Sheva, Israel

RSCPublishing

ISBN: 978-0-85404-155-8

A catalogue record for this book is available from the British Library

Published by The Royal Society of Chemistry,
Thomas Graham House, Science Park, Milton Road,
Cambridge CB4 0WF, UK

Registered Charity Number 207890

For further information see our web site at www.rsc.org

αc

Preface

Vibrationally mediated photodissociation (VMP) deals with the influence of vibrational excitation of the ground electronic state of a molecule on its dissociation following excitation of this state to a higher electronic state. VMP studies were initially motivated by the prospect of enhancing the photodissociation cross section as a result of vibrational pre-excitation. However, more recently the main interest in VMP has been driven by the hope that by preparing some specific vibrational levels rather than others, the subsequent dissociation of the electronically excited molecule could be controlled. In other words, the goal of achieving so-called "mode-selective" or "bond-selective" dissociation has been at the heart of VMP studies. Obviously, for this goal the molecule should have at least two vibrational modes. Therefore, the focus of VMP studies has been on triatomic or larger molecules.

The scope of this monograph is to present the methodology of VMP *via* state-of-the-art specific examples. Overviews of earlier works are included as well, to serve as a background for current research. The presentation deals with "isolated" neutral molecules in the gas phase; VMP in complexes, clusters or condensed phases is not discussed. Wherever appropriate, original studies are quoted, including the original drawings. It is hoped that from the description of the motivation, the approach, the execution of the experiment and the analysis of the results of the specific examples, the reader will get a comprehensive understanding of the field. In addition, a general introduction and summary should give the necessary orientation before and after plunging into details. The limited length of the monograph imposes selection of only a limited number of examples from the extensive literature dealing with VMP that could be discussed, but an effort has been made to give references to as many studies as we were aware of. We apologize for any possible oversights.

The level of presentation is appropriate for senior undergraduate and graduate students of chemistry and physics. It is hoped that the monograph will serve as an introduction to VMP for beginners and as a literature guide to those acquainted with the subject but not necessarily working on VMP. Obviously, in

Vibrationally Mediated Photodissociation
By Salman (Zamik) Rosenwaks
© Salman (Zamik) Rosenwaks 2009
Published by the Royal Society of Chemistry, www.rsc.org

a monograph of limited length, very detailed explanations of the experimental and theoretical methods involved in VMP could not always be included; however, references to original studies are presented at the end of each chapter and should help readers who are interested in more details.

I would like to thank the students of our laboratory and colleagues all over the world for their learned input *via* very many discussions we have had in meetings and through the mail. During the writing of the monograph I have contacted many authors of the papers cited in the monograph and I apologize for not mentioning them all, since a full list would include tens of names. I would still like to mention those whose help has been particularly extensive, our graduate and postgraduate students Amir Golan, Chen Levi, Ran Marom and Alex Portnov and my friends and colleagues Ilana Bar, who also deserves most of the credit for the VMP work conducted in our laboratory, Paul Dagdigian, Aryeh Levin and Moshe Shapiro. Last but not least, I am greatly indebted to the members of my family, whose names I am not mentioning at their request. Without their support and patience, this monograph would not have been written.

Salman (Zamik) Rosenwaks

Contents

Abbreviations and Acronyms		xiii
Useful Physical Constants and Conversion Factors		xiv
Chapter 1	**Introduction**	**1**
	1.1 What is VMP and How Does it Work?	1
	1.2 Why VMP?	4
	1.3 Organization of the Monograph	6
	References	7
Chapter 2	**Theoretical Aspects**	**9**
	2.1 Photodissociation Dynamics	9
	2.1.1 Potential-Energy Surfaces, the Born–Oppenheimer Approximation and the Franck–Condon Principle	10
	2.1.2 Photodissociation Cross Sections: Computational Approaches	13
	2.1.2.1 Classical Trajectories	14
	2.1.2.2 Time-Independent Approach	14
	2.1.2.3 Time-Dependent Approach	15
	2.1.2.4 Comparison of the Time-Independent and Time-Dependent Approaches	18
	2.1.3 Radiationless Transitions between PESs, the Role of Conical Intersections	18
	2.1.4 Vector Correlations in VMP	20
	2.2 Intramolecular Vibrational Dynamics: Normal and Local Modes and the Role of Intramolecular Vibrational Redistribution	21

Vibrationally Mediated Photodissociation
By Salman (Zamik) Rosenwaks
© Salman (Zamik) Rosenwaks 2009
Published by the Royal Society of Chemistry, www.rsc.org

	2.2.1	Normal Modes and Local Modes	22
	2.2.2	Intramolecular Vibrational Redistribution	23
	References		27

Chapter 3   Experimental Methods   **29**

3.1	Preparation and Detection of Vibrational States	32
	3.1.1 Nonspecific Vibrational Excitation	32
	3.1.2 One-Photon, State-Selected Vibrational Excitation	32
	3.1.3 Two-Photon, State-Selected Vibrational Excitation	33
	3.1.3.1 Overtone–Overtone Double Resonance	33
	3.1.3.2 Preparation by Stimulated Raman Excitation and Detection by Coherent Anti-Stokes Raman Spectroscopy and Photoacoustic Raman Spectroscopy	34
	3.1.3.3 Preparation by Stimulated Emission Pumping	35
	3.1.3.4 Preparation by Stimulated Raman Adiabatic Passage	35
	3.1.4 Monitoring Vibrational Excitation	36
3.2	Excitation and Detection of Electronic States	36
3.3	Detection and Characterization of Photofragments	37
	3.3.1 Spontaneous Emission and Laser-Induced Fluorescence	37
	3.3.2 Resonantly Enhanced Multiphoton Ionization	37
	3.3.3 Doppler Profiles	38
	3.3.4 High-n Rydberg Time-of-Flight	38
	3.3.5 Photofragment Translational Spectroscopy	39
	3.3.6 Velocity-Map Imaging	39
	References	40

Chapter 4   VMP of Diatomic Molecules and Radicals   **43**

4.1	HF	44
4.2	HCl	44
4.3	HBr	45
4.4	HI	46
4.5	OH, OD, SH and SD Radicals	47
4.6	O_2	48
	References	48

Chapter 5   VMP of Triatomic Molecules Excluding Water   51

    5.1   VMP of O_3 51
        5.1.1 Spectral Features of O_3 Relevant to VMP   51
        5.1.2 VMP of O_3 as a Source of $O(^1D)$ 53
        5.1.3 Vector Correlations in the VMP of O_3 54
    5.2   VMP of OCS   55
        5.2.1 Spectral Features of OCS Relevant
               to VMP   55
        5.2.2 VMP of OCS in the Second and Higher
               Absorption Bands   55
        5.2.3 VMP of OCS in the First Absorption Band   57
        5.2.4 Vector Correlations in the VMP of OCS   59
    5.3   OCSe   60
    5.4 N_2O 60
    5.5 CS_2 61
    5.6   ICN   63
    5.7   HCN, DCN, HOCl, DOCl and HOBr   63
    References   64

Chapter 6   VMP of Water Isotopologues   68

    6.1   Spectral and Dynamical Features of Water Relevant
         to VMP   68
    6.2   VMP of H_2O and HOD in the Second Absorption
         Band   70
    6.3   VMP of H_2O in the First Absorption Band   71
        6.3.1 Vibrational-State Dependence of H_2O
               Absorption Cross Section   71
        6.3.2 Vibrational Distributions of the OH
               Photofragments   73
        6.3.3 Rotational, Spin-Orbit and Λ-Doublet State
               Distributions of the OH Photofragments   75
            6.3.3.1 Rotational Distributions   77
            6.3.3.2 Spin-Orbit and Λ-Doublet
                    Distributions   78
        6.3.4 Vector Correlations in the VMP of H_2O 79
        6.3.5 VMP of H_2O as a Spectroscopic Tool   79
    6.4   VMP of HOD   81
        6.4.1 VMP of HOD Pre-Excited to Fundamental
               OH or OD Vibrations   82
        6.4.2 VMP of HOD Pre-Excited to OH Overtones   84
        6.4.3 VMP of HOD Pre-Excited to OD Overtones   88
    6.5   VMP of D_2O 90
    References   90

Chapter 7   VMP of Tetratomic Molecules   **94**

    7.1   VMP of Acetylene Isotopologues   94
        7.1.1 Spectral and Dynamical Features of Acetylene
              Relevant to VMP   94
        7.1.2 VMP of C_2H_2 96
            7.1.2.1 Adiabatic VMP of C_2H_2 97
                (a) The $4v_{CH}$ Region   97
                (b) Comparison of VMP for Different
                Vibrational Levels in the $5v_{CH}$ Region   97
                (c) Rotational Effects in the $5v_{CH}$
                Region   101
                (d) Comparison of VMP for Different
                Vibrational Levels in the $4v_{CH}$ Region   103
                (e) The $1v_{CH}$ Region   104
            7.1.2.2 Nonadiabatic VMP of C_2H_2 104
        7.1.3 VMP of C_2HD 107
            7.1.3.1 Adiabatic VMP of C_2HD – Evidence
                for Rotational Dependence of H/D
                Branching Ratio   107
            7.1.3.2 Nonadiabatic VMP of C_2HD 110
    7.2   VMP of Ammonia Isotopologues   110
        7.2.1 Spectral Features of Ammonia Relevant
              to VMP   110
        7.2.2 VMP of NH_3 111
            7.2.2.1 VMP of NH_3 Prepared at Low
                Vibrational Energies   111
            7.2.2.2 VMP of NH_3 Prepared at High
                Vibrational Energies   116
            7.2.2.3 VMP of NH_3 as a Spectroscopic
                Tool   117
        7.2.3 VMP of NHD_2 and NH_2D 118
    7.3   VMP of HNCO   120
        7.3.1 Spectral Features of HNCO Relevant
              to VMP   120
        7.3.2 A Brief Survey of the VMP of HNCO   121
        7.3.3 Comparison of VMP of HNCO $3v_1$ to
             One-Photon Dissociation of Thermal
             Molecules   123
        7.3.4 Application of VMP to HNCO
             Spectroscopy   128
    7.4   VMP of Other Tetratomic Molecules   129
        7.4.1 Hydrogen Peroxide Isotopologues   131
        7.4.2 Nitrous Acid   134
        7.4.3 Hydrazoic Acid   135
    References   136

Chapter 8   VMP of Larger than Tetratomic Molecules   142

8.1   VMP of Methylamine Isotopologues   143
   8.1.1   Spectral and Dynamical Features Relevant
           to VMP of Methylamine   143
   8.1.2   VMP of CH_3NH_2 in the Region of the
           Fundamental CH_3 and NH_2 Stretches   145
       8.1.2.1   Evidence for Mode-Dependent
                 Enhancement of Bond Fission and
                 Photoionization of CH_3NH_2 146
       8.1.2.2   H-Atom Doppler Profiles: Evidence
                 on the Main Photodissociation
                 Channel   148
   8.1.3   VMP of CH_3NH_2 in the Region of the NH_2
           Overtone Stretches   149
   8.1.4   VMP of CD_3NH_2 150
8.2   VMP of Haloalkanes   151
   8.2.1   Hydrochlorofluorocarbons   151
       8.2.1.1 $CHFCl_2$ 152
       (a) Enhanced Production of
           Photofragments   152
       (b) Evidence for Vibrationally Induced
           Three-Body Photodissociation   153
       8.2.1.2 CH_3CFCl_2 156
       (a) Enhanced Production of
           Photofragments   156
       (b) The Effect of Vibrational Pre-Excitation
           on the Branching Ratios of the Atomic
           Photofragments in the VMP of CH_3CFCl_2
           and CH_3CF_2Cl 156
   8.2.2   Methyl Iodide   157
       8.2.2.1   Photodissociation in the First
                 Absorption Band: General
                 Background   157
       8.2.2.2   Photodissociation of Vibrationally
                 Excited CH_3I 159
   8.2.3   Other Haloalkanes   162
       8.2.3.1 CF_3I 163
       8.2.3.2 CH_3Cl and CHD_2Cl 163
       8.2.3.3 CH_2Cl_2 164
8.3   VMP of Phenol   164
8.4   VMP of Other Larger than Tetratomic Molecules   166
   8.4.1 $HONO_2$, $(CH_3)_3COOH$, CH_3OH, NH_2OH,
           HO_2NO_2 and CH_3OOH 168
       8.4.1.1 $HONO_2$ ns and fs-ps VMP
                 Studies   168

	8.4.1.2	$(CH_3)_3COOH$	169
	8.4.1.3	CH_3OH	169
	8.4.1.4	NH_2OH	170
	8.4.1.5	HO_2NO_2	171
	8.4.1.6	CH_3OOH	172
8.4.2	Acetylene Homologues		172
	8.4.2.1	Photodissociation Dynamics	173
		(a) $D_3CC{\equiv}CH$	173
		(b) $H_3CC{\equiv}CH$	174
		(c) $H_3CH_2CC{\equiv}CH$	174
	8.4.2.2	Applications to Vibrational Spectroscopy	176
8.4.3	Ethene Isotopologues		177
	8.4.3.1	The Fourth C–H Stretching Overtone Region in $H_2C{=}CH_2$, *Trans*-HDC=CDH and $H_2C{=}CD_2$	178
	8.4.3.2	Vibrational Patterns of the First Through Fourth C–H Stretching Overtone Regions in $H_2C{=}CH_2$	179
	8.4.3.3	The First Stretch Overtone Region in *Trans*-HDC=CDH: Example of VMP-Based Detailed Analysis of Vibrational Spectroscopy and Dynamics	180
8.4.4	UF_6		183
References			183

Concluding Remarks   189

Author Index   192

Subject Index   201

Abbreviations and Acronyms

BO	Born–Oppenheimer
CARS	Coherent anti-Stokes Raman spectroscopy (or scattering)
CI	Conical intersection
DFT	Density functional theory
FC	Franck–Condon
fs	Femtosecond
HRTOF	High-n Rydberg time-of-flight
IC	Internal conversion
IR	Infrared
IVR	Intramolecular vibrational redistribution
LIF	Laser-induced fluorescence
LM	Local mode
μm	Micrometer
Nd:YAG	Neodymium-doped yttrium aluminum garnet; $Nd:Y_3Al_5O_{12}$
NIR	Near-infrared
nm	Nanometer
NM	Normal mode
ns	Nanosecond
OPO	Optical parametric oscillator
PA	Photoacoustic
PARS	Photoacoustic Raman spectroscopy (or scattering)
PES	Potential-energy surface
ps	Picosecond
PTS	Photofragment translational spectroscopy
REMPI	Resonantly enhanced multiphoton ionization
RRKM	Rice–Ramsperger–Kassel–Marcus
SEP	Stimulated emission pumping
SRE	Stimulated Raman excitation
STIRAP	Stimulated Raman adiabatic passage
TOFMS	Time-of-flight mass spectrometer
UV	Ultraviolet
VIS	Visible
VMI	Velocity-map imaging
VMP	Vibrationally mediated photodissociation
ZOBS	Zero-order bright state
ZODS	Zero-order dark state

Useful Physical Constants and Conversion Factors

Table 1 Physical constants (rounded). Symbols for units: $C =$ Coulomb, $J =$ Joule, $K =$ Kelvin.

Quantity	Symbol	Value	SI Units	Cgs Units
Electron charge	e	1.6022	10^{-19} C	10^{-20} emu
Electron mass	m_e	9.1095	10^{-31} kg	10^{-28} g
Bohr radius (atomic units)	a_0	5.2918	10^{-11} m	10^{-9} cm
Rydberg constant	R_∞	1.09737	10^7 m^{-1}	10^5 cm^{-1}
Speed of light in vacuum	c	2.99792	10^8 m s^{-1}	10^{10} cm s^{-1}
Planck's constant	h	6.6261	10^{-34} Js	10^{-27} erg s
$h/2\pi$	$\hbar$	1.05459	10^{-34} Js	10^{-27} erg s
Gas constant	R	8.3144	10^3 J kmol^{-1} K^{-1}	10^7 erg mol^{-1} K^{-1}
Avogadro's number	N_A	6.0221	10^{26} kmol^{-1}	10^{23} mol^{-1}
Boltzmann's constant (R/N_A)	k	1.38066	10^{-23} J K^{-1}	10^{-16} erg K^{-1}

Table 2 General conversion factors.

Length
1 Ångström (Å) $= 10^{-10}$ m $= 10^{-1}$ nm; 1 micrometer (μm) $= 10^{-6}$ m
Force
1 Newton (N) $= 1$ kg m s^{-2} $= 10^5$ dyn $= 10^5$ g cm s^{-2}
Pressure
1 Pascal (Pa) $= 1$ N m^{-2} $= 10^{-5}$ bar $= 10$ dyn cm^{-2}
1.01325×10^5 Pa $= 1$ atm $= 1.01325 \times 10^6$ dyn cm^{-2} $= 760$ torr
Energy
1 Joule (J) $= 1$ kg m^2 s^{-2} $= 10^7$ erg $= 0.239$ cal
1 Electronvolt (eV) $= 96.485$ kJ mol^{-1}

Table 3 Energy conversion factors. Numbers in parentheses denote powers of 10 by which the entry is to be multiplied.

	erg	J	cal	eV	cm^{-1}	Hz	K	kJ mol^{-1}	kcal mol^{-1}
1 erg =	1	1.0000(−7)	2.390(−8)	6.241(+11)	5.034(+15)	1.5092(+26)	7.243(+15)	6.022(+13)	1.4393(+13)
1 J =	1.0000(+7)	1	2.390(−1)	6.241(+18)	5.034(+22)	1.5092(+33)	7.243(+22)	6.022(+20)	1.4393(+20)
1 cal =	4.1840(+7)	4.1840	1	2.611(+19)	2.106(+23)	6.315(+33)	3.031(+23)	2.520(+21)	6.022(+20)
1 eV =	1.6022(−12)	1.6022(−19)	3.829(−20)	1	8.0651(+3)	2.418(+14)	1.1605(+4)	9.648(+1)	2.306(+1)
1 cm^{-1} =	1.9865(−16)	1.9865(−23)	4.748(−24)	1.2399(−4)	1	2.998(+10)	1.4388	1.1963(−2)	2.859(−3)
1 Hz =	6.626(−27)	6.626(−34)	1.5837(−34)	4.136(−15)	3.336(−11)	1	4.799(−11)	3.990(−13)	9.537(−14)
1 K =	1.3807(−16)	1.38071(−23)	3.300(−24)	8.6171(−5)	6.950(−1)	2.084(+10)	1	8.314(−3)	1.9871(−3)
1 kJ mol^{-1} =	1.6606(−14)	1.6606(−21)	3.969(−22)	1.0364(−2)	8.359(+1)	2.506(+12)	1.2027(+2)	1	2.390(−1)
1 kcal mol^{-1} =	6.948(−14)	6.948(−21)	1.6606(−21)	4.3371(−2)	3.498(+2)	1.0486(+13)	5.032(+2)	4.184	1

CHAPTER 1

Introduction

1.1 What is VMP and How Does it Work?

Vibrationally mediated photodissociation (VMP) is a special case of photo-dissociation where *vibrational* excitation of (in most cases) the ground electronic state of a molecule precedes its *electronic* excitation to a higher, dissociative electronic state.[1-6] In a sense, all photodissociation processes conducted in a sample of molecules could be considered as VMP due to thermal vibrational population, unless $kT \ll$ lowest vibrational frequency of the molecules. However, although we will occasionally refer to examples of thermal vibrational population, our main interest will be state-selected VMP, where a specific rovibrational state is prepared. State-selected VMP utilizes, mostly, pulsed lasers in a double-resonance scheme where both the vibrational pre-excitation and the electronic excitation are resonant. As noted in the preface, we will deal with "isolated" neutral molecules in the gas phase; VMP in complexes, clusters or the condensed phase will not be discussed. VMP studies aim to unravel the influence of the vibrational pre-excitation on the photodissociation cross section, the quantum states of the photofragments and the branching ratio between them as well as their angular distribution. This influence is discussed in detail in the following chapters. Here, we introduce the concept of VMP *via* simple examples. The methodology of state-selected VMP is exemplified in Figure 1.1 for a bound–free transition (excitation to a purely repulsive electronic state) in a molecule that contains at least one hydrogen atom, RH, where R represents one or more atoms. RH is chosen as an example since this class of molecules has been the most popular VMP species due to the R–H vibrational modes that are readily excited (at higher frequencies than other modes). Further details on the experimental techniques utilized in the three steps described in the figure, vibrational pre-excitation, excitation to a dissociative electronic state and detection of the ensuing photofragments, as well as on other techniques, are given in Chapter 3.

Vibrationally Mediated Photodissociation
By Salman (Zamik) Rosenwaks
© Salman (Zamik) Rosenwaks 2009
Published by the Royal Society of Chemistry, www.rsc.org

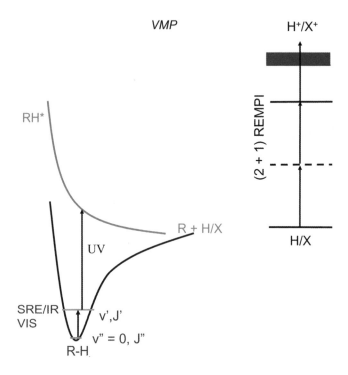

Figure 1.1   An artist's concept of the methodology of state-selected VMP in an RH molecule. It consists of three steps: (1) An R–H vibrational mode is initially excited to a specific (v', J') state; an infrared (IR) photon or stimulated Raman excitation (SRE) is utilized for preparing fundamental vibrations or low overtones, near-IR (NIR) or visible (VIS) photon for exciting high overtones. (2) An ultraviolet, UV, (or sometimes VIS) photon is utilized for electronic excitation and UV photons for (3) monitoring the ensuing photofragments (H or X, which is, in many VMP studies, a halogen atom). In the present example resonantly enhanced multiphoton ionization (REMPI) is shown as the means for monitoring an atomic photofragment; the solid line denotes the resonant state reached by two-photon excitation and the dashed line an intermediate, virtual state. Molecular fragments are commonly monitored *via* laser-induced fluorescence (LIF).

A schematic description of the potential-energy curves and absorption spectra in "non-VMP" and VMP is depicted in Figures 1.2 (a) and (b), respectively, for a bound–free transition in a diatomic molecule – photodissociation from the vibrationless ground state (a) and from a vibrationally excited state (b) are compared. The former is characterized by a broad structureless absorption spectrum of the parent molecule, due to the reflection of the vibrationless ground-state wavefunction on the upper potential-energy curve. The latter is characterized by a spectrum that reflects extended regions on the ground-state potential onto the dissociative upper state.[2] Figure 1.2 indicates that when the dissociation

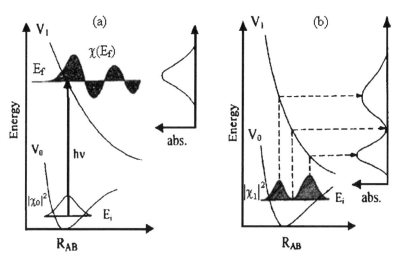

Figure 1.2 Schematic illustration of photodissociation of a diatomic molecule AB *via* excitation to a purely repulsive electronic state. V_0 and V_1 are the potential-energy curves for the ground (bound) and excited (repulsive) electronic states, respectively, as a function of the internuclear distance R_{AB}, E_i and $E_f = E_i + E_{photon}(= h\nu)$ are the initial and final energies of these states. (a) Absorption from the vibrationless ground state χ_0 to $\chi(E_f)$, the continuum wavefunction. (b) Absorption from the $v = 1$ vibrationally excited state of the ground electronic state, χ_1: transitions from the mid-point and from locations close to the turning points of the classical vibration are shown (the shapes of the different $\chi(E_f)$ functions are omitted here). Promotion of extended regions on the ground electronic state onto the dissociative potential-energy curve is reflected in the absorption spectrum in both (a) and (b). Adapted from Ref. 4 with permission.

is promoted from a vibrationally excited state, different regions of the repulsive excited state are accessed compared to those from the vibrationless ground state.

It is very important to note here that the transitions shown in Figures 1.1 and 1.2 are vertical, namely, the coordinates of the nuclei in the molecule remain unchanged during the electronic transition. This is not fortuitous, it exemplifies the Franck–Condon (FC) principle that plays a major role in VMP and is discussed and referred to throughout the monograph. Indeed, as we will see, VMP is effective when there is a favorable FC overlap between the vibrationally excited states of the ground electronic state and those of the electronically excited dissociative or predissociative states. If, in addition, the initially prepared vibrational states survive intramolecular vibrational redistribution (IVR) until the subsequent electronic excitation takes place, dissociation of the electronically excited molecule could be controlled by preparing specific vibrational states.

The direct, bound–free photodissociation of a diatomic molecule presented in Figure 1.2(b) is the simplest example of VMP. However, the major part of the monograph deals with larger molecules and, in many cases, with indirect photodissociation, where a potential barrier in the dissociation exit channel precludes direct dissociation. In molecules larger than diatomic, multi-dimensional potential-energy surfaces (PES) determine the motion of the atoms. The potential-energy curves of a diatomic molecule (Figure 1.2) also represent one-dimensional cuts through a multidimensional PES and often serve as a simple description of VMP in larger molecules.

In most cases VMP has been carried out in the first absorption band of the molecule at hand. This is since in the studied molecules absorption from vibrationally excited states of this band is usually in the mid- and near-ultraviolet (UV) or visible (VIS) wavelength regions readily accessed by existing high-power lasers. $\lambda \sim 193$ nm was popular in the early days of VMP due to the availability of the ArF excimer laser; longer wavelengths have been mostly applied in later studies. Also, almost all state-selected VMP studies have been performed with pulsed lasers with a temporal pulsewidth of at least a few nanoseconds (ns). This was partly due to the fact that these lasers provided the required energy for VMP – typically millijoules and microjoules for vibrational and electronic excitation, respectively; ns pulses also enabled selective rovibrational excitation with higher resolution than that possible with shorter pulses.

Numerous methods have been devised for studying a variety of VMP aspects and they are described when individual molecules are discussed in the following chapters. To conclude this introduction on how VMP works, we mention two noteworthy points: (1) VMP is often compared to isoenergetic one-photon excitation from the vibrationless ground electronic state. For this comparison the wavelength for electronic excitation in the former has to be redshifted with respect to that in the latter (see Figure 1.2). (2) Many recent VMP studies have been conducted in samples cooled *via* supersonic expansion. This results in reducing the Doppler linewidths of the monitored transitions and the rotational temperature as well as the population of low-lying vibrational states (and thus minimizes the contribution of hot bands to the spectrum). All this leads to less congestion and extra narrowing and simplification of the spectrum.

1.2   Why VMP?

VMP has been of both theoretical and applied interest. From a theoretical perspective, exploring how vibrational pre-excitation affects the course of photodissociation and correlating the findings with the topology of PES have long been important parts of molecular processes studies.[2,5,6] As we will see in many examples, one-photon dissociation and VMP, conducted at the same total energy, often result in different quantum states and branching ratios of the photofragments. This means that vibrational pre-excitation may have a dynamic role rather than an energetic role. The role of the FC factors in the efficacy of the electronic excitation step and of the exit-channel interactions in the dissociation process have been extensively studied and referred to in the examples presented in the monograph. The most prominent conclusion inferred from these studies is that due to the FC principle, starting the electronic excitation from a specific vibrationally excited state rather than another, enables access to a different region of the upper PES and may affect the dissociation.

The theoretical investigations have been accompanied by experimental studies aimed at applications of VMP for several purposes. Most notable is the application of VMP to the so-called "mode-selective" or "bond-selective" dissociation, namely, to control the dissociation products (the terms mode-selective and bond-selective are usually used as synonyms although, strictly speaking, the former should refer to selective pre-excitation of a vibrational mode and the latter to the resulting selective bond cleavage). This aspect of VMP will be extensively presented in many examples.

A related but somewhat different application of VMP is for isotope separation. Isotope separation that is based on the different frequencies of electronic excitation of different isotopes of atoms or molecules depends on the small difference in the electron reduced mass when the atoms are light and on nuclear volume effects (*e.g.*, core polarization) when the atoms are heavy.[7] However, this isotopic shift is much smaller than the shift in the vibrational frequencies in molecules due to the large difference in the reduced mass of the vibrating oscillator.[8] It is thus easier to selectively excite a specific rovibration of a specific molecular isotope (so-called isotopologue) and then photodissociate it rather than other isotopologues by using appropriate redshifted wavelengths. Moreover, by preparing different vibrational states in a molecule that contains different atomic isotopes, different isotopic photoproducts can be preferentially produced from the same molecule (this is a special case of the above-mentioned control of dissociation products). The pre-eminent example of preferential production of different isotopes from the same molecule is the VMP of HOD (see Section 6.4).

An additional application of VMP is in vibrational spectroscopy. Monitoring high-overtone and combination transitions encounters the problem that their intensities are exponentially decreasing functions of the total change of vibrational quantum number, Δv. A crude rule of thumb is that the magnitude of this decrease is $\sim 10^{-\Delta v}$ for hydride stretches.[9] Also, when spectroscopy studies are conducted in molecular beams, the low density of the target molecules adds to the difficulty of measurement.

These difficulties can be largely alleviated by applying VMP for the following reasons. (1) The action spectrum, *i.e.* the photoproduct yield as a function of the wavelength of the vibrational pre-excitation, is monitored rather than directly measuring the intensity of vibrational bands. The appearance of the bands in the action spectrum is often enhanced due to a better FC overlap of the initially prepared vibrational state in the ground electronic state with that in the excited state. (2) Even if the FC overlap is not favorable, the detection of the photoproduct by resonantly enhanced multiphoton ionization (REMPI) adds several orders of magnitude to the signal as compared to direct detection of vibrational bands, since ions can be monitored with very high sensitivity. Signal enhancement is also achieved, although to a lesser extent, for detection by laser-induced fluorescence (LIF).

The role of VMP in other molecular spectroscopy and molecular dynamics scenarios will be recognized from the examples given in the text. A notable example, with important implications to atmospheric chemistry, is production of the atmospherically important species $O(^1D)$ from vibrationally excited ozone in the longer-wavelength "tail" ($\lambda > 310\,\text{nm}$) of its solar photolysis. Indeed, VMP may play a role in any environment where molecules are electronically excited.

1.3   Organization of the Monograph

The methodology is that VMP is presented in this monograph *via* specific examples, starting with diatomic molecules (Chapter 4), continuing with tri- and tetratomic molecules (Chapters 5–7) and ending with larger molecules (Chapter 8). Original studies are quoted, wherever appropriate, and many original drawings are adapted. Due to the limited length of the monograph, a choice has to be made. Some examples, which were considered heuristic, are presented in detail, others are only briefly described or just cited. Once again, we apologize for any possible oversight. To become acquainted with the technical language used in these examples, two short overviews precede Chapters 4–8, one on theoretical aspects and one on experimental methods.

Chapter 2 lists the key questions that have to be addressed for understanding VMP, which include, but go beyond, those addressed in photodissociation in general. It then explains the main approaches for calculating PESs and in this context presents the Born–Oppenheimer approximation, which is also the basis for the FC principle. Approaches for computing photodissociation cross sections are then presented, namely, classical trajectories, and quantum-mechanical time-independent and time-dependent approaches; this is followed by an explanation of the concept of conical intersections, which is utilized to account for radiationless transitions between nonadiabatic PESs. The next part of this chapter briefly discusses vector correlations in VMP, *i.e.* the relation between the directions of the vectors inherent in the initiation and dynamics of the photodissociation process, when the photons applied for both vibrational pre-excitation and electronic excitation are polarized. In the final part of the chapter intramolecular vibrational dynamics, in particular IVR, is discussed.

In Chapter 3 we present the major experimental methods for VMP studies. The experimental realization of VMP involves three major steps: Preparation of a vibrational state, promotion of this state to a higher PES(s) followed by dissociation of the molecule, and finally interrogation of the ensuing photo-fragments. The main experimental approaches usually utilized for each step are presented. It is noteworthy that for vibrational preparation we present methods that have been widely used as well as some that have been hardly (or not yet) applied in VMP but have a potential for exciting vibrational levels that are otherwise difficult to reach.

In Chapter 4 VMP of diatomic molecules is discussed, only briefly, since mode- or bond-selective dissociation, which has been the main reason for the recent interest in VMP, is not relevant for these molecules. However, the presentation of VMP studies of diatomic molecules serves to demonstrate, for the simplest cases, the effect of rovibrational parent excitation on the photodissociation dynamics, *e.g.*, enhancement of the dissociation cross section, polarization dependence of product angular distribution, and branching ratios between electronic states of the products.

Chapters 5 and 6 discuss the VMP of triatomic molecules, the simplest polyatomics where mode- or bond-selective dissociation can be demonstrated. Also, triatomics are small enough to allow *ab initio* calculations of PESs and photodynamics, although they possess several vibrational degrees of freedom, like stretches and bends, which play a principal role also in larger molecules. VMP of several triatomic molecules, excluding water, is described in Chapter 5. Chapter 6 presents, in some detail, the VMP studies of water isotopologues. It was the extensive theoretical and experimental investigations of H_2O and HOD that opened a new era of detailed studies of state-to-state photodissociation out of specific rovibrationally excited states of polyatomic molecules.

In Chapter 7 we move on to tetratomic molecules and dwell, in particular, on acetylene isotopologues, ammonia isotoplogues and isocyanic acid, for which extensive state-to-state VMP studies have been carried out. The VMP theories and experiments for tetratomic molecules are obviously more complex than those for triatomics. The additional atom adds additional degrees of freedom that complicate both theory and experiment. Moreover, due to the additional degrees of freedom, IVR is expected to obscure mode selectivity. Also, in contrast to the triatomic molecules, in particular the water molecule where detailed calculations predicted selectivity, similar predictions were not available prior to VMP experiments. Nevertheless, mode selectivity has been observed in several tetratomic molecules.

In Chapter 8 we deal with VMP of even larger molecules, where a much higher complexity is expected. Moreover, it is anticipated that IVR in these molecules would preclude mode- or bond-selective dissociation on the ns timescale. Indeed, at present mode-dependent enhancement in photodissociation has been observed for only one "larger than tetratomic" molecule, methylamine. We start this chapter with a presentation of the VMP of methylamine isotopologues and then move on to haloalkanes, where, in some instances, vibrational pre-excitation changes the electronically excited states accessed in photodissociation and the branching ratio between the atomic photofragments. We then deal with phenol, where the dynamics of conical intersections in this large molecule is exemplified, and finally report on VMP of other "larger than tetratomic" molecules.

References

1. F. F. Crim, *Annu. Rev. Phys. Chem.*, 1993, **44**, 397.
2. R. Schinke, *Photodissociation Dynamics*, Cambridge University Press, Cambridge, 1993.

3. F. F. Crim, *J. Phys. Chem.*, 1996, **100**, 12725.
4. I. Bar and S. Rosenwaks, *Int. Rev. Phys. Chem.*, 2001, **20**, 711.
5. M. Shapiro and P. Brumer, *Principles of the Quantum Control of Molecular Processes*, Wiley, New York, 2003.
6. R. D. Levine, *Molecular Reaction Dynamics*, Cambridge University Press, Cambridge, 2005.
7. L. D. Landau and E. M. Lifshitz, *Quantum Mechanics – Non-Relativistic Theory*, Pergamon Press, Oxford, 1958.
8. G. Herzberg, *Molecular Spectra and Molecular Structure I. Spectra of Diatomic Molecules*, Van Nostrand, Princeton, 1950.
9. D. J. Nesbitt and R. W. Field, *J. Phys. Chem.*, 1996, **100**, 12735.

CHAPTER 2
Theoretical Aspects

Theoretical studies of photodissociation dynamics have been the subject of numerous publications, especially since the 1960s. We list below a few references[1-9] (review articles, books or chapters in books) where fundamental theories of photodissociation dynamics are presented and previous studies in this field are cited. Since VMP is a special case of photodissociation in general, the above studies are the basis for its understanding. In the following sections we briefly describe some aspects of these studies, in particular those relevant to VMP, in order to get acquainted with the technical language used in the monograph.

2.1 Photodissociation Dynamics

The theoretical treatment described below assumes an electric dipole moment induced, weak-field photodissociation with long duration (not less than a few ns) of an exciting light pulse. The dimensions of the photodissociated molecule are (obviously) small compared to the wavelengths of the applied radiation. Under these assumptions, first-order perturbation theory is applicable.[7,8] It is also noteworthy that in most VMP studies a ns timescale is utilized for the initial preparation of vibrational states, the subsequent photodissociation and the delay between the two steps, and the working pressure is low (usually much below 1 Torr). Therefore, the radiative and collisional relaxation of the vibrationally excited molecules can be neglected.

As mentioned above, the main interest in VMP has been in triatomic or larger molecules. A polyatomic molecule is denoted here as ABC, where A, B or C represent one or more atoms. However, since photodissociation theories have dealt more frequently with and are more developed for triatomic molecules than for larger ones, we will refer hereafter to A, B and C as atoms unless otherwise stated.

Vibrationally Mediated Photodissociation
By Salman (Zamik) Rosenwaks
© Salman (Zamik) Rosenwaks 2009
Published by the Royal Society of Chemistry, www.rsc.org

If energetically allowed, *i.e.* if the energy of the initially excited rovibrational state plus the subsequent electronic excitation energy $h\nu$ exceeds the dissociation energy of the ruptured bond, the VMP of ABC can be presented as

$$ABC^{\dagger} + h\nu \rightarrow ABC^{*} \rightarrow A + BC \tag{2.1}$$

$$\rightarrow B + AC \tag{2.2}$$

$$\rightarrow C + AB \tag{2.3}$$

$$\rightarrow A + B + C \tag{2.4}$$

where † and * denote rovibrationally and electronically excited states, respectively. The fragments may be internally (electronically, vibrationally and rotationally) excited, as well as translationally hot. Key questions that have to be addressed for understanding VMP are:

1. What is the IVR lifetime of $ABC^{\dagger}$?
2. What is the cross section for the $ABC^{\dagger} + h\nu \rightarrow ABC^{*}$ process?
3. What is the lifetime of ABC^{*} and how does this lifetime depend on the nature of the PES and on the excitation energy? We note that the lifetime is also a function of the exciting pulse profile.[9]
4. What affects the branching ratio between the energetically accessible channels? If more than one bond breaks, does this occur in a concerted or a sequential process?
5. At a given excitation energy, is the dissociation governed by one electronic state or does the fragmentation take place on several PESs?
6. How is the total available energy distributed among the internal and translational degrees of freedom of the fragments?
7. What is the angular distribution of the fragments?

The answers to most of these questions are principally determined by the shape of the PES of the ground and electronically excited state, and we therefore turn, first, to discuss their features.

2.1.1 Potential-Energy Surfaces, the Born–Oppenheimer Approximation and the Franck–Condon Principle

The common notion of a potential-energy surface, PES,[3,7,8,10–12] that determines the motion of the nuclei, depends on the validity of the Born–Oppenheimer (BO) approximation, where the nuclear and electronic motions are separated. The approximation is justified as long as the motions of the (heavy) nuclei take place on much longer timescales than those of the (light) electrons. It is noteworthy here that the same argument is also at the heart of the Franck–Condon (FC) principle. Namely, during an electronic transition between two states, the coordinates and the corresponding momenta of the nuclei in the molecule remain unchanged; the transition is "vertical" to the so-called FC point. As we will see below, the assumption of vertical transitions is the starting point in the calculations of photodissociation cross sections.

To implement the BO approximation, we present the molecular Hamiltonian as

$$H_{mol}(\mathbf{q}, \mathbf{Q}) = T_{nuc}(\mathbf{Q}) + H_{el}(\mathbf{q}, \mathbf{Q}) \tag{2.5}$$

where $T_{nuc}(\mathbf{Q})$ is the nuclear kinetic energy operator, $H_{el}(\mathbf{q},\mathbf{Q})$ the electronic Hamiltonian (comprising the electronic kinetic energy and all the Coulombic interactions), $\mathbf{q}$ are the coordinates of the electrons and $\mathbf{Q}$ those of the nuclei. The *electronic* (time-independent) Schrödinger equation is solved with the nuclei held fixed (at a geometry denoted by $\mathbf{Q}$), *i.e.* T_{nuc} is ignored in the Hamiltonian in formulating this equation:

$$[H_{el}(\mathbf{q}, \mathbf{Q}) - E(\mathbf{Q})]\psi(\mathbf{q}, \mathbf{Q}) = 0 \tag{2.6}$$

The so-called adiabatic (namely, where no change in quantum numbers takes place) PESs, $E_k(\mathbf{Q})$, and the adiabatic electronic states $\psi_k(\mathbf{q},\mathbf{Q})$ are the eigenvalues and eigenfunctions of the electronic Hamiltonian $H_{el}(\mathbf{q},\mathbf{Q})$. These eigenvalues and eigenfunctions, as well as the Hamiltonian, depend parametrically on $\mathbf{Q}$.

The total, time-independent molecular wavefunction in the adiabatic representation is expanded as

$$\Psi(\mathbf{q}, \mathbf{Q}) = \Sigma_k \chi_k(\mathbf{Q})\psi_k(\mathbf{q}, \mathbf{Q}) \tag{2.7}$$

where $\chi_k(\mathbf{Q})$ are the nuclear wavefunctions. They can be found by utilizing the approximation that the electronic wavefunction varies negligibly with the nuclear coordinates. The *nuclear* Schrödinger equation for the nuclear motion within the kth electronic state, where the electronic potential energies are the $E_k(\mathbf{Q})$ eigenvalue of (2.6), is then:

$$[T_{nuc}(\mathbf{Q}) + E_k(\mathbf{Q}) - E]\chi_k(\mathbf{Q}) = 0 \tag{2.8}$$

We note that in eqn (2.8) there is no coupling of the kth electronic state to other states, meaning that the motion of the nuclei in this state is not affected by nuclear motions in other states, which is the gist of the BO approximation. The Hamiltonian in eqn (2.8) is diagonalized in this approximation[13] and the time-independent molecular wavefunction for the kth electronic state is thus presented as

$$\Psi_k(\mathbf{q}, \mathbf{Q}) = \chi_k(\mathbf{Q})\psi_k(\mathbf{q}, \mathbf{Q}) \tag{2.9}$$

When the nuclei are allowed to move, *e.g.*, near avoided crossings or conical intersections (see below) where large coupling between the PESs exists, the BO approximation is not valid. A diabatic representation, where the nuclear kinetic energy rather than the electronic Hamiltonian is diagonalized,[13] is then more appropriate. We note that in many VMP studies, adiabatic PESs, where conical intersections describe nonadiabatic transitions between them (see below), are the preferred choice of representation. For further discussion of the validity of the BO separation and of adiabatic *vs.* diabatic representation of PES see, *e.g.*, Refs 7,12 and (in particular) 13.

It is noteworthy that *ab initio* calculations of PES (*i.e.* solving the electronic Schrödinger equation) might be very difficult for large systems. Therefore, several methods have been developed to overcome this difficulty. For example, semiempirical methods, where, *e.g.*, interpolations between known "valleys" of the parent molecule and the fragments have been utilized. To facilitate dynamics calculations, PESs obtained applying semiempirical as well as the above-mentioned *ab initio* calculations have often been fitted to analytical functions. Modern methods for constructing PES include: (1) An *ab initio* multiple spawning method that solves the electronic and nuclear Schrödinger equations simultaneously; this makes *ab initio* multiple dynamics approaches applicable for problems where quantum-mechanical effects of both electrons and nuclei are important.[14] (2) Construction of approximate *ab initio* PES of the parent molecule by interpolating the corresponding *ab initio* surfaces of small molecular fragments.[15,16]

The shape of the excited PES determines whether the photodissociation is direct, namely, when the PES is purely repulsive, or indirect, *e.g.*, when there is a potential barrier in the exit channel. An example of the former is the photodissociation of H_2O in the first absorption band (see Chapter 6) and of the latter is that of FNO (Section 2.1.2.3).

For an ABC molecule, the (adiabatic) PES represents the interaction energies in a given electronic state as a function of the configuration of (the nuclei of) the A, B and C atoms, *e.g.*, of the three interatomic distances (for an N-atomic molecule the PES is a function of the $3N$-6 internal degrees of freedom), or alternatively of two distances and the angle between them or of one distance and two angles. If A, B or C are molecules, additional parameters have to be included. Since multidimensional PESs are difficult to visualize, the PESs of polyatomic molecules are graphically presented as one-dimensional or two-dimensional cuts through a multidimensional PES. In the one-dimensional presentation the interaction is given either as a function of the distance between two of the atoms (which is in fact the potential-energy curve for a diatomic molecule), or between one atom and the center of mass of the others, where the other distances are frozen. Alternatively, it is given as a function of the angle between the distances or the torsion angle of one of the bonds. In the two-dimensional presentation the interaction is given either as a function of two distances (sometimes as a contour map) or of a distance and an angle.

An example of PES of an electronically excited state is presented in Figure 2.1 where a two-dimensional cut through the six-dimensional PES of $HNCO(\tilde{A}\,^1A'')$ is depicted as a function of the two bond distances R_{HN} and R_{NC}. The two exit channels presented in the figure correspond to the two chemically different

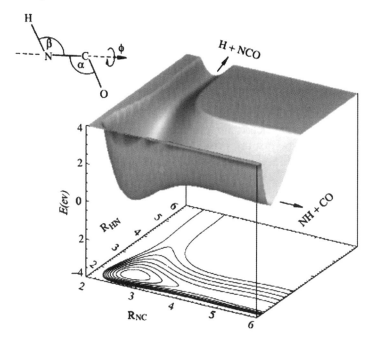

Figure 2.1 Two-dimensional representation of the potential-energy surface for the $\tilde{A}$ $^1A'' \leftarrow$ state of HNCO as a function of the NC and the HN bond distances R_{NC} and R_{HN}, respectively. The other coordinates are frozen. Reproduced from Ref. 8 with permission.

products $H + NCO$ and $NH + CO$ that can be formed (as shown in Section 7.3, the photodissociation of HNCO leads to *three* distinct channels, since NH can be formed in both triplet, $NH(X^3\Sigma^-)$, and singlet, $NH(a\ ^1\Delta)$, states). Although the other four degrees of freedom, R_{CO}, α; β and ϕ, are fixed in this picture, the potential is not independent of them and a complete dynamics calculation should include all of them.

2.1.2 Photodissociation Cross Sections: Computational Approaches

In this section we provide a brief overview of the main computational approaches for photodissociation cross sections. A general approach unifying the different quantum-mechanical methods is presented in Refs 4 and 9. However, we follow hereafter, Sections 2.1.2.1–4, the presentation of Refs 7 and 8, which are the closest ones to that utilized in many papers reporting on VMP experiments. We start by outlining the classical trajectory approach and then describe the time-independent and time-dependent quantum-mechanical approaches. It is important to note that although, as shown below, the last two approaches provide different views and computational methods, they are completely equivalent and yield the same cross sections[4,7,8,9,17] We elaborate on this point in Section 2.1.2.4.

2.1.2.1 Classical Trajectories

As mentioned above, if the nuclei move much slower than the electrons the FC principle holds and the electronic transition from the initial to the final excited state is assumed to be instantaneous. Thus, the internal (nuclear) coordinates and the corresponding momenta of the molecule in its initial state remain unchanged during the excitation. In the classical approach to photodissociation, the movements of the nuclei, after promotion to the PES of the excited state, are governed by the Hamilton equations of motion[18]

$$\frac{dQ_i}{dt} = \frac{\partial H}{\partial P_i}, \quad \frac{dP_i}{dt} = -\frac{\partial H}{\partial Q_i} \tag{2.10}$$

where $H(\mathbf{Q},\mathbf{P})$ is the Hamilton function, the classical analogue of the quantum-mechanical operator H in the excited state, $\mathbf{Q}(t)$ represents the nuclear coordinates and $\mathbf{P}(t)$ the corresponding momenta. The equations of motion are solved with the initial values $\mathbf{Q}(0)$ and $\mathbf{P}(0)$, weighted to the distributions in the initial state and the calculation of the photodissociation cross section is averaged over the $(\mathbf{Q},\mathbf{P})$ phase space. This approach can be picturesquely described as placing "billiard balls" on the excited PESs, near the FC point (vertically above the initial point in the lower PES) and following their trajectories towards different exit channels.

The classical approach is useful, in particular, for large systems where quantum-mechanical calculations are difficult. However, semiclassical methods that incorporate quantum mechanics are utilized if quantum-mechanical effects like interferences have to be accounted for.

2.1.2.2 Time-Independent Approach

Under the assumptions listed at the beginning of Section 2.1, we can represent the photodissociation cross sections by simple expressions. To further simplify the presentation we denote the above time-independent wavefunctions of eqn (2.9), where we add the index i for the initial state and f for the final state with energy $E_i + E_{photon}$, as follows: $\chi_{k,i}(\mathbf{Q}) \equiv n$, $\chi_{k,f}(\mathbf{Q}) \equiv n'$, $\psi_{k,i}(\mathbf{q},\mathbf{Q}) \equiv e$ and $\psi_{k,f}(\mathbf{q},\mathbf{Q}) \equiv e'$. For $E_{photon} = h\nu$, the partial photodissociation cross section for producing the fragments in a particular state α is given by Fermi's Golden Rule[18a]

$$\sigma(\nu, \alpha) \sim |\langle n(\alpha)'|\langle e'|\mu_e|e\rangle|n\rangle|^2 \tag{2.11}$$

where μ_e is the component of transition electronic dipole moment operator in the direction of the polarization of the excitation electric field vector. We note that for continuum wavefunctions in the final state, the coordinates accounting for the separation of the photofragments extend to infinity. From the BO approximation eqn (2.11) can be presented as

$$\sigma(\nu, \alpha) \sim |\langle n(\alpha)'|n\rangle|^2 |\langle e'|\mu_e|e\rangle|^2 \tag{2.12}$$

The total photodissociation cross section, *i.e.* the absorption cross section, is obtained by summing the partial cross sections over all α quantum numbers

$$\sigma(\nu) = \sum_{\alpha} \sigma(\nu, \alpha) \tag{2.13}$$

To obtain the VMP signal we need to know, in addition, the expression for the cross section for vibrational pre-excitation of the initial electronic state, which is given by

$$\sigma(v) \sim |\langle n|\mu_v|g\rangle|^2 \tag{2.14}$$

where $|g\rangle$ is the ground vibrational state of the initial electronic state and μ_v is the analogue of μ_e for the vibrational transition. The VMP signal can thus be presented by

$$S(\nu, \alpha, v) \sim |\langle n(\alpha)'|n\rangle|^2 |\langle e'|\mu_e|e\rangle|^2 |\langle n|\mu_v|g\rangle|^2 \tag{2.15}$$

In simple words, eqn (2.15) means that the VMP signal depends on the FC overlap between the nuclear wavefunctions in the final and initial electronic states, $|\langle n(\alpha)'|n\rangle|^2$, on the electronic transition dipole moment and on the vibrational excitation probability in the initial electronic state. The VMP signal presented in eqn (2.15) is the quantity usually measured in VMP experiments.

It is important to note that in the above simplified presentation we ignored the rotational component of the nuclear wavefunction. The effect of rotational FC factors is discussed in Refs. 7 and 19. We also note that FC models are not always sufficient to account for the observed VMP spectra and exit-channel interactions play a significant role in the photodissociation dynamics of some molecules, notably H_2O. The role of rotational FC factors and of exit-channel interactions is discussed in the chapters presenting VMP of specific molecules.

2.1.2.3 Time-Dependent Approach

As in the time-independent approach, we assume here the conditions listed at the beginning of Section 2.1, namely, weak-field photodissociation with long duration of an excited light pulse and small dimensions of the photodissociated molecule. The time-dependent approach relies on solving the nuclear time-dependent Schrödinger equation

$$\left[i\hbar \frac{\partial}{\partial t} - H_f \right] \Phi_f(\mathbf{Q}, t) = 0 \tag{2.16}$$

where the index f denotes the final, excited state and Φ_f the time-dependent wavepacket evolving $[\Phi_f(0) \to \Phi_f(t)]$ on the excited PES. The time-dependent approach resembles the classical trajectories approach in being an initial value problem. Also, as in that approach, the electronic transition from the initial to the final excited state is assumed to be fast, the nuclear coordinates in the initial state remaining unchanged.

$\Phi_f(0)$ is obtained from the relation

$$\Phi_f(0) = M_{fi}\chi_i(E_i) \tag{2.17}$$

where $M_{fi} = \langle e'|\mu_e|e \rangle$ and $\chi_i(E_i)$ is the initial (ground) bound-state wavefunction with energy E_i. We note that for finite time length pulses (certainly for ns or longer pulses), $\Phi_f(0)$ is made up of a sequence of replicas of $\chi_i(E_i)$ (multiplied by M_{fi}) launched by ultrashort pulses on the excited PES at different times and multiplied by the pulse amplitude at these times. However, this does not affect the conclusions reached using the calculation procedure presented here.[17]

Since $\Phi_f(0)$ is not an eigenstate of H_f, it becomes a time-evolving wavepacket on the excited PES. Its propagation on this PES is obtained by solving eqn (2.16):

$$\left| \Phi_f(t) \right\rangle = e^{-iH_f t/\hbar} \left| \Phi_f(0) \right\rangle \tag{2.18}$$

where $e^{-iH_f t/\hbar}$ is the time-evolution operator. The overlap of the evolving wavepacket at time t with that at $t = 0$ is the autocorrelation function

$$S(t) = \langle \Phi_f(0) | \Phi_f(t) \rangle \tag{2.19}$$

The total absorption cross section is then[7]

$$\sigma(E_f) \sim \int_{-\infty}^{+\infty} e^{-iE_f t/\hbar} S(t) \mathrm{d}t \tag{2.20}$$

The energy-dependent cross section is thus the Fourier transform of the time-dependent autocorrelation function. The partial photodissociation cross sections are obtained by projecting the wavepacket $\Phi_f(t)$ for $t \to \infty$ on the stationary wavefunctions of the asymptotic Hamiltonian (where for $t \to \infty$ the interaction potential between the departing species vanishes).[7]

An illuminating illustration of the motion of a wavepacket is presented in Figure 2.2, which shows, in a two-dimensional coordinate space, snapshots of

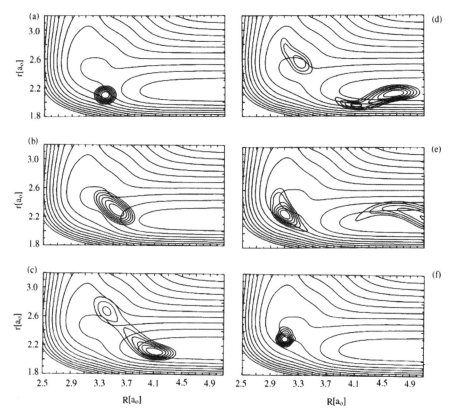

Figure 2.2 Snapshots of the evolving wavepacket in the S_1 excited electronic state of FNO. $\Phi_f(R,r,t)$ is plotted for various times in fs: (a) 0; (b) 7.26; (c) 14.52; (d) 21.78; (e) 29.04; and (f) 36.30. R and r are "Jacobi coordinates": R is the distance from the recoiling F atoms to the center of mass of NO and r the N–O separation (given in atomic units). Reprinted with permission from Ref. 20. Copyright 1992, American Institute of Physics.

the evolving wavepacket for the photodissociation of FNO in the S_1 state.[20] The initial wavepacket is prepared at the FC point, *i.e.* at the equilibrium coordinates of the ground electronic state. For $t < 10$ femtosecond (fs), Figure 2.2(b), the wavepacket follows the steepest descent and slides down the slope towards the tiny potential barrier, just like a classical "billiard ball". At ~ 14 fs, near the barrier, Figure 2.2(c), it starts to split into two parts; then one portion enters the exit channel and slides down the steep potential trough leading to fragments F and NO in a time much shorter than an internal vibrational period, Figures 2.2(d) and (e). The other part is trapped in the inner region of the potential ridge, performs one full oscillation, and returns to its starting position, Figure 2.2(f). Here, a new round begins, another portion of the wavepacket leads to dissociation, whereas the remaining wavepacket performs a second oscillation. This continues until the entire wavepacket has finally escaped from the inner region and travels freely in the F + NO exit channel. This trapping and oscillatory behavior of the wavepacket is a typical example of indirect photodissociation. Without a barrier, *i.e.*, for a purely repulsive PES, the photodissociation would be direct with the *entire* wavepacket rapidly moving to the exit channel, without coming back to its starting position.

2.1.2.4 Comparison of the Time-Independent and Time-Dependent Approaches

As mentioned above, the two approaches are completely equivalent and yield the same cross sections. The equivalence is inherent in the presentation of photodissociation in Refs. 4,9 and 17, and explained in detail in Ref. 7, which is followed here. The equivalence is demonstrated by considering the stationary and time-dependent wavefunctions presented above. The total dissociation wavefunction $\chi_t(E_f)$ is defined as the sum over all α states of the partial dissociation amplitude (at the final energy E_f) multiplied by the final dissociation function $\chi_f(E_f, \alpha)$:

$$\chi_t(E_f) = \sum_\alpha |\langle n(\alpha)'|\langle e'|\mu_e|e\rangle|n\rangle|\chi_f(E_f, \alpha) \qquad (2.21)$$

As shown in Ref. 7, $\Phi_f(t)$ can be expanded as a coherent superposition of the $\chi_t(E_f, \alpha)$ stationary states, each being multiplied by the time-evolution function $e^{-iH_f t/\hbar}$. It then follows that

$$\chi_t(E_f) = \int_{-\infty}^{+\infty} e^{-iE_f t/\hbar}\, \Phi_f(t)\mathrm{d}t \qquad (2.22)$$

namely, $\chi_t(E_f)$, which comprises all dissociation processes for the energy E_f, is the Fourier transform of the time-dependent wavepacket $\Phi_f(t)$, which consists of all stationary states for all energies.

Throughout the monograph examples of the utilization of the two approaches are presented, the choice between the two being dependent on the particular system. One advantage of the time-independent approach is that it directly computes the overlap elements in eqn. (2.15), enabling a clear picture of VMP processes to be obtained. An advantage of the time-dependent approach is that information for all energies can be extracted from a single wavepacket.

2.1.3 Radiationless Transitions between PESs, the Role of Conical Intersections

In the previous sections we described processes where excitation to a PES is followed by dissociation on this excited surface. If the surface is not purely repulsive and there is a potential barrier on the way to dissociation, tunneling through the barrier may occur. Another possibility is that a nonadiabatic

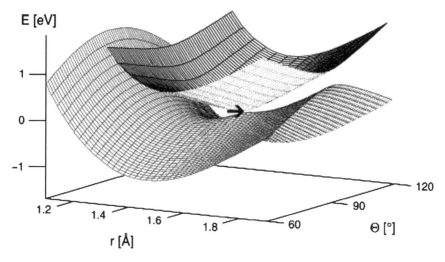

Figure 2.3 An example of adiabatic S_0 and S_1 surfaces in the vicinity of a conical intersection (CI) of a molecule containing a C=C bond. r is the bond distance and θ the torsion angle around the bond. The CI point in this example is at $r^0_{CI}=1.59$ Å, $\theta = 90°$ and is marked by an arrow. Reproduced from Ref. 22 by permission of the Royal Society of Chemistry.

radiationless transition from the initially excited PES to another PES, with subsequent dissociation, takes place (a radiationless transition may also follow tunneling between two valleys on the excited state). Such a transition is classified as internal conversion when no change in total spin is involved and as an intersystem crossing when a change in total spin occurs. These transitions are usually strongest when induced by conical intersections (CI).[21–24]

A CI of two PESs (Figure 2.3) is the set of molecular geometry points where the two intersect (*i.e.* are degenerate). The efficacy of CIs between PES 1 and 2 is governed by the derivative couplings, projected gradients of the involved electronic wavefunctions, ψ_1 and ψ_2. The couplings between the adiabatic electronic states are given by $\langle \psi_1 | \frac{\partial}{\partial Q_j} \psi_2 \rangle$ and higher derivatives with respect to the nuclear coordinates Q_j. It is the finite range of these couplings that extends the "effective size" of the CIs to a non-negligible volume and is thus responsible for nonadiabatic events.

Additional features of CIs are worth noting. At a CI adiabaticity breaks down – nuclei would have to change directions "instantaneously" and electrons cannot keep up. A CI requires that the adiabatic electronic wavefunction changes sign when transported along a closed loop in nuclear coordinate space surrounding the region of the intersection. The change in sign in the *electronic* wavefunction requires a compensating change in the sign of the *nuclear* wavefunction, known as the geometric phase (sometimes referred to as the Berry phase) effect (the change of sign means that the phase changes by π which is the common case when the loop encloses one CI; the general rule is that a CI requires a change of $n\pi$, where $n = 1, 2, 3, \ldots$). It is also noteworthy that the radiationless transitions mentioned above can proceed without CIs; this occurrence depends on the size of nonadiabatic coupling, on the nuclear dynamics and on the occurrence of the geometric phase effect.

By their particular topology, CIs act as "photochemical funnels" that effectively and selectively induce transitions between electronic states. They are a consequence of the BO separation of nuclear and electronic motion and represent an essential aspect of electronically nonadiabatic processes. Figure 2.3 is a schematic presentation of an example of a CI between two singlet adiabatic PESs, the ground S_0 and the excited S_1, for a molecule that contains a C=C bond.[22] The PESs are depicted as a function of the C=C distance r and the torsion angle θ around the C=C bond, the other bonds and angles being frozen.

2.1.4 Vector Correlations in VMP

Vector correlations[25] refer to the relation between the directions of the vectors inherent in the initiation and dynamics of the photodissociation process: The polarization of the electric vector $\mathbf{E}$ of the light beam (or beams in VMP) acting on the parent molecule, the parent transition dipole moment $\boldsymbol{\mu}$, the relative velocities of the recoiling fragments, $\mathbf{v}$, and the rotational angular momentum of the molecular photofragment(s) and the orbital angular momentum polarization of the atomic photofragment(s), both denoted by $\mathbf{J}$. The correlations involving $\mathbf{E}$ or $\boldsymbol{\mu}$ (preferentially directed parallel to $\mathbf{E}$), relate to the laboratory frame, whereas the $\mathbf{v}$–$\mathbf{J}$ correlation relates only to the body frame and thus most directly reflects exit-channel interactions.

A frequently asked question is where do the velocities of the recoiling fragments point with respect to the direction of the linear polarization of the light beam employed for the dissociation. Indeed, the $\boldsymbol{\mu}$–$\mathbf{v}$ correlation, the angular scattering distribution of the recoiling photofragments, $I(\theta)$, is encountered in many photodissociation studies, including VMP. For a one-photon photodissociation event using linearly polarized light this is given by[25,26]

$$I(\theta) = \frac{\sigma}{4\pi}[1 + \beta \cdot P_2(\cos\theta)] \tag{2.23}$$

where θ is the angle between $\mathbf{v}$ and $\mathbf{E}$, σ the partial cross section for the dissociation product, $P_2(x)$ the second-order Legendre polynomial ($P_2(x) = (3x^2 - 1)/2$) and β is the anisotropy parameter. The magnitude and sign of β are related to the orientation of the transition dipole moment, $\boldsymbol{\mu}$, in the parent molecule, the symmetry of the excited state and the excited-state lifetime. Within the axial recoil approximation, $\beta = -1$ corresponds to a perpendicular transition, whereas $\beta = +2$ corresponds to a parallel transition. Another correlation reported in some VMP studies is $\boldsymbol{\mu}$–$\mathbf{J}$, the rotational alignment. The limiting values of the alignment, denoted by $A_0^{(2)}$, at the high-J limit are[27,28] 0.8 for $\boldsymbol{\mu} \parallel \mathbf{J}$ and -0.4 for $\boldsymbol{\mu} \perp \mathbf{J}$. The $\mathbf{v}$–$\mathbf{J}$ (for both rotational and orbital angular momenta) and μ–v–J correlations are also encountered in VMP studies. As we will see later in some specific examples of VMP, common methods utilized to unravel various correlations are probing the LIF spectra and the Doppler profiles of the fragments produced in the photodissociation.

Like one-photon dissociation (assumed in the above description), which depends on the polarization of the photon, VMP is an anisotropic process. Moreover, it may depend on the polarization of both the photons applied for vibrational pre-excitation and for electronic excitation. The theory of vector correlations in VMP has been treated by employing classical trajectories,[29] semiclassical[30] and quantum-mechanical approaches.[31,32] In the classical approach,[29] the vibrational influence on the anisotropy parameter β in photo-dissociation of triatomic molecules is explored. The calculations indicate that in a fast dissociation, accompanied by a large angular momentum excitation of the diatomic fragment, the recoil direction can change significantly, resulting in a measurable modification of β.[29] Employing the semiclassical approach, a general expression for the Doppler profile for fragments produced in the photo-dissociation of laser excited, aligned molecules, as in VMP, is derived.[30] The scope of the quantum-mechanical analysis of vector correlations in VMP is different in Refs. 31 and 32. In the former the interest is in controlling the atomic polarization in the dissociation of diatomic molecules. In the latter, Doppler-shift-dependent fluxes for various arrangements of the polarization directions of the vibrational excitation and photolysis lasers are computed. It is found that the profiles depend upon the photodissociation dynamics only through the conventional recoil anisotropy parameter β, as in one-photon dissociation. The main conclusion of the analysis of this work[32] is that, generally, the dynamical information obtainable from measurements of vector correlations in VMP is similar to that extractable in one-photon dissociation. Nevertheless, in the VMP case the observables also depend on additional geometrical factors due to the alignment of the excited intermediate states and additional information can be provided on the dissociation dynamics since different regions of PES are accessed.

2.2 Intramolecular Vibrational Dynamics: Normal and Local Modes and the Role of Intramolecular Vibrational Redistribution

Since the starting point of VMP is vibrational pre-excitation of the initial (usually ground) electronic state, understanding the intramolecular vibrational dynamics in this state is crucial for understanding VMP. The vibrational dynamics on the excited electronic state reached as a consequence of electronic excitation from the initially prepared state is important as well. However, the vibrational dynamics of the ground electronic state is more frequently addressed in VMP studies.

The most important concept in this context is that of intramolecular vibrational redistribution, IVR, which may cause the vibrational excitation of a specific mode to spread quickly and in an apparently complex manner over the entire molecular framework. The theory of IVR has been extensively studied. We list here a few publications,[33–39] mostly feature or review articles. However, in our concise presentation below we follow, in particular, Refs. 35,36 and 37, since these references include pedagogical overviews of IVR that fit well our goal to briefly review theoretical aspects that are relevant to VMP.

2.2.1 Normal Modes and Local Modes

To facilitate the discussion of IVR, we first recapitulate the formalism for describing vibrations in polyatomic molecules. In the description in the framework of normal modes (NM), the vibrations are treated as small oscillations of the nuclei within the vicinity of the equilibrium configuration in a harmonic potential.[18,40,41] Under this assumption of "pure" harmonic vibrations, each atom in the molecule is moving with the same frequency and phase for a given NM. In terms of the normal coordinates along which the motions of NM occur, both the potential and kinetic energies in the Hamiltonian are diagonal (*i.e.*, of quadratic form). However, to describe the vibrational states in a real molecule, in particular high-lying states, coupling of several NMs has to be assumed. This deviation from the pure harmonic approximation is generally referred to as anharmonicity. *The anharmonicities are introduced as perturbations to the wave equation*, *e.g.*, cubic terms and terms containing products that involve different normal coordinates are added to the potential-energy expression. The NM models utilized in vibrational spectroscopy include these couplings. The energies of the vibrational levels for a polyatomic molecule can then be obtained from the expression[40]

$$
G(v_1, v_2, \ldots) = \sum_i \omega_i \left(v_i + \frac{d_i}{2} \right) + \sum_i \sum_{j \geq i} x_{ij} \left(v_i + \frac{d_i}{2} \right) \left(v_j + \frac{d_j}{2} \right)
$$
$$
+ \sum_i \sum_{j \geq i} g_{ij} l_i l_j + \ldots \tag{2.24}
$$

where v_i is the vibrational quantum number, ω_i the harmonic vibration wave number for the NM i, d_i the corresponding degeneracy, and x_{ij} and g_{ij} are anharmonicity constants, where the latter contributes only to the energy of degenerate vibrations with vibrational angular momentum quantum numbers l_i and l_j.

The NM framework is less suitable for treating high-lying molecular vibrations, in particular those that have relatively small couplings to other bonds, *e.g.*, stretching vibrations of bonds to light atoms (notable examples are X–H bonds where X is O, C or N). These vibrations are better described in the framework of the local-mode (LM) model,[42–44] where the vibrations are related to motions of *individual anharmonic bonds* (often approximated by Morse oscillators). Coupling among the bonds is not included, although addition of interbond coupling may be required for better LM description. Still, anharmonicity is an integral part of the LM model, whereas interbond coupling is inherent in the NM model. In the LM model the energy of i equivalent Morse oscillators with n_i quanta in the i-th oscillator is expressed as[45]

$$G(n_1, n_2, ...) = \sum_i \left[\omega_m \left(n_i + \frac{1}{2} \right) + x_m \left(n_i + \frac{1}{2} \right)^2 \right] \quad (2.25)$$

where ω_m is the harmonic wave number and x_m the anharmonicity of the Morse oscillator.

Expansion in either LM or NM zero-order basis describes the same molecular eigenstates of the full Hamiltonian. The choice of the model depends on the vibrational states being investigated and in case of doubt a rule of thumb is to choose the one in which further correction terms (anharmonicities for NM or interbond couplings for LM) are the smallest. A useful approach is to employ a hybrid LM/NM model that treats different vibrations by the different models.[46] For example, in Ref. 47 LM basis functions were used for C–H(D) stretching vibrations and NM basis functions to describe C–H(D) bending vibrations in $CH_3(D_3)X$ molecules (X denotes a halogen atom).

When examining a molecule prepared in a highly vibrationally excited state by photon absorption, we can benefit from the fact that the optically active state often corresponds to a simple vibrational motion close to one of the zero-order states (of either the LM or NM basis). This optically active state is referred to as a *zero-order bright state* (ZOBS).[35–37] At sufficiently high energy, where the density of states is large, there will be many other states that carry little or no oscillator strength (at least in comparison with the ZOBS), and these are referred to as *zero-order dark states* (ZODS). *Although called "states", the ZOBS and ZODS are not molecular eigenstates of the full Hamiltonian.* However, both types of states are eigenstates of the zero-order Hamiltonian that neglects anharmonic coupling terms. For the full molecular Hamiltonian that includes these coupling terms, the resulting eigenstates can be expressed as linear combinations of the ZOBS and ZODS:

$$|n\rangle = C_b^n |b\rangle + \sum_d C_d^n |d\rangle \quad (2.26)$$

where $|b\rangle$ denotes a ZOBS and $|d\rangle$ a ZODS. Indeed, eqn (2.26) reflects the mixing between the zero-order states at high vibrational energies, which is substantial when *anharmonic* terms in the potential become significant. Figure 2.4 displays a schematic view of the ZOBS (first column) interacting with a set of ZODSs (second column) to form molecular eigenstates $|n\rangle$ (third column), which are a linear combinations of the bright and dark states. The absorption intensity of each transition (last column) is proportional to the fraction of ZOBS mixed into the particular vibrational eigenstate.

2.2.2 Intramolecular Vibrational Redistribution

In addition to displaying the above states, Figure 2.4 presents a simple interpretation of IVR in the frequency domain. A single rovibrational ZOBS is

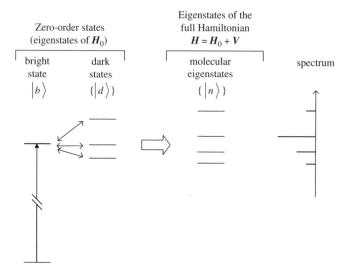

Figure 2.4   Schematic description of zero-order states and the molecular eigenstates formed by their interaction. In this simple model for IVR, the full Hamiltonian is separated into a zero-order term (H_0) and an interaction term (V). The zero-order bright state (*e.g.*, the stretch state) carries all of the oscillator strength from the ground state. This bright state is embedded in a bath of optically inactive dark states and is coupled to them by the anharmonic terms initially neglected. Once these terms are included in the full Hamiltonian, the resulting eigenstates are mixtures of the zero-order states and each can carry some brightness. In the spectrum resulting from scanning the frequency of excitation (*e.g.*, by a laser), each eigenstate appears with intensity proportional to its brightness. Adapted from Ref. 37 by permission of the Royal Society of Chemistry.

assumed to carry all of the oscillator strength in the excitation energy region. This ZOBS is coupled to the near-resonant ZODSs through perturbations (V) in the zero-order Hamiltonian (H_0). This coupling results in molecular eigenstates that mix the character of the ZOBS and ZODSs. As a result, the ZOBS oscillator strength is distributed among these states. The width of the distribution of intensity is a measure of the timescale of energy localization in the bright state. The center of gravity of the "IVR multiplet" is preserved at the original position of the bright state.

Anharmonic coupling between zero-order vibrational states is often exhibited by a Fermi-resonance interaction, a notable example being that between the symmetric stretch and the two bendings of the CO_2 molecule.[40] In addition to the strong low-order Fermi resonance interactions, the spectrum of highly excited molecules shows extensive local perturbations (resulting from high-order anharmonics, Darling–Dennison resonances, Coriolis or centrifugal couplings).[40] Therefore, if the spectrum is measured with sufficient resolution, sets of many transitions can be observed. The minimal resolution that allows the lines of an IVR multiplet to be resolved should be 1/(number of states per unit energy). If the experimental resolution is larger than this limit, the IVR multiplet will not be resolved and will appear as a single homogeneously broadened rovibrational line with the width given by Fermi's Golden Rule:[18a]

$$\Gamma = 2\pi\langle V^2 \rangle \rho \tag{2.27}$$

where Γ is the linewidth, ρ is the bath state density and V is the root-mean-squared interaction matrix element of the bright state with the bath states. It should be noted that this width, which is proportional to the density of states, is actually a rate of single-exponential decay of the population in the ZOBS (in time-domain experiments). The "characteristic" IVR lifetime is given by $1/\Gamma$; however, as explained below, the IVR lifetime may depend on a variety of molecular properties.

To illustrate the effects of higher-order perturbations, IVR is frequently described in terms of *tiers* of coupled zero-order states (Figure 2.5).[36] At the coarsest level of the structure is the ZOBS itself. As the spectral resolution is increased, the oscillator strength associated with the ZOBS fractionates into finer states, sometimes called "feature" or "doorway" states. As one proceeds systematically from lower to higher resolution, one is able to see more and more states coupled to the ZOBS by progressively weaker and weaker off-diagonal matrix elements of the molecular Hamiltonian (expressed in the zero-order basis set).

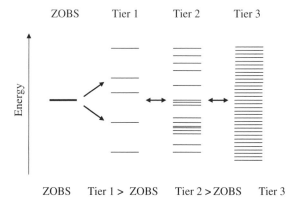

Figure 2.5 Tier model for analysis of IVR. The IVR is described in terms of *tiers* of coupled zero-order states. Energy typically spreads from an initially localized state *via* strong, low-order anharmonic interactions with a few near-degenerate basis states. These states are in turn each coupled to a successive tier of near-degenerate basis states *via* additional low-order resonances. The net effect is a hierarchical coupling scheme, sequential energy flow, and a progressive set of tiers with increasing density of states and decreasing coupling with the ZOBS. The states in tier 1 are denoted as "doorway" states. Reprinted with permission from Ref. 36. Copyright 1996 American Chemical Society.

IVR occurs in the time domain but, as indicated above, can be investigated in the frequency domain by employing various combinations of supersonic cooling and high-resolution lasers. Analysis of linewidths in unresolved spectra on the one hand and of eigenstate-resolved spectra on the other hand in a variety of molecules has revealed wide ranges of timescales for IVR. In sufficiently small molecules, the timescale can be extremely long, up to a few ns, allowing selectively excited vibrations to influence the outcome of photo-dissociation. On the other hand, certain functional groups exhibit ultrafast energy redistribution on the timescale of fs. Two limiting cases of coupling between vibrational states and the relation to their lifetime (τ) or splitting can be represented as follows.[48] If the number of coupled states is large, energy in the initially excited vibrational state will decay exponentially (and irreversibly) into the dense bath states and the spectral transition will appear in the frequency domain as a smooth envelope with width (in units of cm^{-1}) $\Delta v = 1/2(\pi c\tau)$. If, on the other hand, the initially excited vibrational state is strongly coupled to only one other vibrational state that is weakly coupled to the rest of the vibrational states, a periodic return of the vibrational energy to the excited state will be observed. This will result in a splitting in the frequency domain, where two peaks spaced by $\Delta v = 1/c\tau$ are observed, where τ is the period of recurrence.

From the Hamiltonian obtained from frequency-domain spectra it is possible to extract the temporal behavior of the zero-order states rather than measure it directly *via* time-domain spectroscopy using short (fs) laser pulses. The vibrational Hamiltonian allows us to follow the temporal evolution of the density matrix given as a function of time by[49]

$$P(t) = U(t)P(t=0)U^{\dagger}(t) \tag{2.28}$$

where

$$U = e^{-iHt/\hbar} \tag{2.29}$$

The diagonal element $P_{ii}(t)$ gives the probability for finding the system at time t in the ith zero-order basis state.

An illuminating manifestation of the complex dependence of IVR on a variety of molecular properties (and not necessarily on the density of states) is the multiple timescales observed in IVR of highly excited vibrational states of methanol.[50,51] Preparing the OH-stretch overtones in the range of $v_1 = 3$–8, the observed splittings and widths were found to correspond to three timescales.[50] The largest splittings imply subpicosecond oscillation of energy between the O–H stretch and a combination with the C–H stretch ($5v_1 \leftrightarrow 4v_1 + v_2$ and $6v_1 \leftrightarrow 5v_1 + v_2$) or a combination with the COH bend ($7v_1 \leftrightarrow 6v_1 + 2v_6$). Secondary timescales, in the picosecond (ps) range, correspond to finer splittings and are thought to arise from low-order resonances with other vibrational states. The nonmonotonic energy dependence and the extent of the secondary structure throughout the recorded spectra reflect the requirement of resonance with important zero-order states. The third timescale, represented by the widths of the narrowest features at each overtone level, reflects the onset of vibrational energy randomization. These widths increase exponentially with vibrational energy in the range $2v_1$ to $8v_1$. At the highest energy $(25\,000\,cm^{-1})$ the three timescales begin to converge, implying an irreversible decay of the OH-stretch overtone in a couple of hundred fs. In a later study,[51] conformational dependence of IVR was observed in methanol in the $5v_1$ state, prepared close to the staggered and partially eclipsed conformations.

References

1. K. F. Freed and Y. B. Band, in: *Excited States*, ed. E. Lim, **Vol. 3**, Academic Press, New York, 1978, pp. 109.
2. E. J. Heller, *Acc. Chem. Res.*, 1981, **14**, 368.
3. E. J. Heller, in: *Potential-Energy Surfaces and Dynamics Calculations*, ed. D. G. Truhlar, Plenum, New York, 1981, pp. 103.
4. M. Shapiro and R. Bersohn, *Annu. Rev. Phys. Chem.*, 1982, **33**, 409.
5. G. G. Balint-Kurti and M. Shapiro, *Adv. Chem. Phys.*, 1985, **60**, 403.
6. M. N. R. Ashfold and J. E. Baggott, ed. *Advances in Gas-Phase Photochemistry and Kinetics: Molecular Photodissociation Dynamics*, Royal Society of Chemistry, London, 1987.
7. R. Schinke, *Photodissociation Dynamics*, Cambridge University Press, Cambridge, 1993.
8. R. Schinke, in: *The Encyclopedia of Computational Chemistry*, ed. P. v. R. Schleyer, N. L. Allinger, T. Clark, J. Gasteiger, P. A. Kollman, H. F. Schaefer III and P. R. Schreiner, Wiley, Chichester, 1998, p 2064.
9. M. Shapiro and P. Brumer, *Principles of the Quantum Control of Molecular Processes*, Wiley, New York, 2003.
10. J. N. Murrell, S. Carter, S. C. Farantos, P. Huxley and A. J. C. Varandas, *Molecular Potential Energy Functions*, Wiley, Chichester, 1984.
11. D. M. Hirst, *Potential-Energy Surfaces*, Taylor and Francis, London, 1985.
12. R. D. Levine, *Molecular Reaction Dynamics*, Cambridge University Press, Cambridge, 2005.
13. H. Köppel, W. Domcke and L. S. Cederbaum, *Adv. Chem. Phys.*, 1984, **57**, 59.
14. M. Ben-Nun and T. J. Martínez, *Adv. Chem. Phys.*, 2002, **121**, 439.
15. M. A. Collins, *Theor. Chem. Acc.*, 2002, **108**, 313.
16. M. A. Collins, *J. Chem. Phys.*, 2007, **127**, 024104.
17. M. Shapiro, *J. Phys. Chem.*, 1993, **97**, 7396.
18. H. Goldstein, *Classical Mechanics*, Addison-Wesley, Reading, 1950.
18(a). C. Cohen-Tannoudji, B. Diu and F. Laloe, *Quantum Mechanics*, Wiley, New York, 1977.
19. G. G. Balint-Kurti, *J. Chem. Phys.*, 1986, **84**, 4443.
20. H. U. Suter, J. R. Huber, M. von Dirke, A. Untch and R. Schinke, *J. Chem. Phys.*, 1992, **96**, 6727.

21. D. R. Yarkony, *J. Phys. Chem. A*, 2001, **105**, 6277.
22. I. Burghardt, L. S. Cederbaum and J. T. Hynes, *Faraday Discuss.*, 2004, **127**, 395.
23. W. Domcke, D. R. Yarkony and H. Köppel, (ed.), *Conical Intersections: Electronic Structure, Dynamics and Spectroscopy*, World Scientific, Singapore, 2004.
24. M. Baer, *Beyond Born–Oppenheimer: Electronic Non-Adiabatic Coupling Terms and Conical Intersections*, Wiley, Hoboken, NJ, 2006.
25. R. N. Zare, *Angular Momentum*, Wiley, New York, 1988.
26. R. N. Zare, *Mol. Photochem.*, 1972, **4**, 1.
27. C. H. Green and R. N. Zare, *Annu. Rev. Phys. Chem.*, 1982, **33**, 119.
28. C. H. Green and R. N. Zare, *J. Chem. Phys.*, 1983, **78**, 6741.
29. H-P. Loock, J. Cao and C. X. W. Qian, *Chem. Phys. Lett.*, 1993, **206**, 422.
30. R. J. Barnes, A. Sinha, P. J. Dagdigian and H. M. Lambert, *J. Chem. Phys.*, 1999, **111**, 151.
31. J. G. Underwood and I. Powis, *J. Chem. Phys.*, 2000, **113**, 7119.
32. P. J. Dagdigian, *J. Chem. Phys.*, 2002, **116**, 7948.
33. M. Bixon and J. Jortner, *J. Chem. Phys.*, 1968, **48**, 715.
34. M. Quack, *Annu. Rev. Phys. Chem.*, 1990, **41**, 839.
35. K. K. Lehmann, G. Scoles and B. H. Pate, *Annu. Rev. Phys. Chem.*, 1994, **45**, 241.
36. D. J. Nesbitt and R. W. Field, *J. Phys. Chem.*, 1996, **100**, 12735.
37. A. Callegari and T. R. Rizzo, *Chem. Soc. Rev.*, 2001, **30**, 214.
38. M. Gruebele and P. G. Wolynes, *Acc. Chem. Res.*, 2004, **37**, 261.
39. D. M. Leitner and M. Gruebele, *Mol. Phys.*, 2008, **106**, 433.
40. G. Herzberg, *Molecular Spectra and Molecular Structure II. Infrared and Raman Spectra of Polyatomic Molecules*, Van Nostrand, Princeton, 1945.
41. E. B. Wilson, J. C. Decius and P. C. Cross, *Molecular Vibrations*, Dover Publications, New York, 1980.
42. B. R. Henry, *Acc. Chem. Res.*, 1977, **10**, 207.
43. M. S. Child and L. Halonen, *Adv. Chem. Phys.*, 1984, **57**, 1.
44. M. S. Child, *Acc. Chem. Res.*, 1985, **18**, 45.
45. I. M. Mills and A. G. Robiette, *Mol. Phys.*, 1985, **56**, 743.
46. M. M. Law and J. L. Duncan, *Mol. Phys.*, 1994, **83**, 757.
47. M. M. Law, *J. Chem. Phys.*, 1999, **111**, 10021.
48. O. V. Boyarkin, R. D. F. Settle and T. R. Rizzo, *Ber. Bunsenges. Phys. Chem.*, 1995, **99**, 504.
49. J. E. Baggott, D. W. Law, P. D. Lightfoot and I. M. Mills, *J. Chem. Phys.*, 1986, **85**, 5414.
50. O. V. Boyarkin, T. R. Rizzo and D. S. Perry, *J. Chem. Phys.*, 1999, **110**, 11346.
51. P. Maksyutenko, O. V. Boyarkin, T. R. Rizzo and D. S. Perry, *J. Chem. Phys.*, 2007, **126**, 044311.

CHAPTER 3
Experimental Methods

A large number of experimental techniques and many different combinations of them have been employed in VMP experiments. The basic scheme consists of three main steps: (1) vibrational pre-excitation, (2) excitation to a dissociative (or predissociative) electronic state and (3) detection of the ensuing photofragments. These steps are described schematically in Figure 1.1 and in more detail in Figure 3.1. As in the former figure, and for the same reasons, an RH molecule serves as an example in the latter. The schemes in the figure describe experiments where a different pulsed laser is used for each step. However, as explained below, in many instances the same laser pulse can be used for steps (2) and (3). The typical temporal width of the laser pulses applied in each step, as well as the delay between steps (1) and (2), is a few ns. In most experiments there is no delay between steps (2) and (3) and this is obviously true when the same laser pulse is used for both steps. We also note that in many experiments LIF rather than REMPI (shown in Figure 3.1) is applied for detecting molecular photofragments.

Panels (a), (b) and (c) in Figure 3.1 describe three types of VMP measurements. In each of them the wavelength of one of the lasers, represented by a thick arrow, is scanned, whereas those of the others, represented by thin arrows, are held fixed (but see below for a possible exception of the latter rule in case (c)). In the measurement described in panel (a), where the same laser pulse can be used for steps (2) and (3), the yield of the ensuing photofragments is monitored as a function of the wavelength of the vibrational excitation and the so-called infrared (IR) action spectrum is obtained. Since this is the most common measurement in VMP studies, we frequently refer to it simply as the action spectrum (note that fundamental vibrations can also be excited *via* stimulated Raman excitation (SRE) and high overtones by VIS photons). In the measurement described in panel (b) the yield of the ensuing photofragments is monitored as a function of the wavelength of the electronic excitation and the UV action spectrum is obtained. In the measurement described in panel (c) the wavelength for vibrational excitation is parked on a specific rovibrational

Vibrationally Mediated Photodissociation
By Salman (Zamik) Rosenwaks
© Salman (Zamik) Rosenwaks 2009
Published by the Royal Society of Chemistry, www.rsc.org

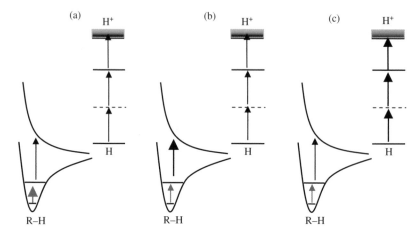

Figure 3.1 The three types of measurements utilized in the basic VMP experiments presented in the text, where REMPI is employed for photofragment detection. In each type of measurement (presented here for the case where a different pulsed laser is used for each excitation step), the wavelength of one of the lasers, represented by a thick arrow, is scanned, whereas those of the others, represented by thin arrows, are held fixed. (a) IR action spectrum: the yield of the ensuing photofragments is monitored as a function of the wavelength of the vibrational excitation (note that fundamental vibrations are also excited *via* SRE and high overtones by VIS photons). (b) UV action spectrum: the yield of the ensuing photofragments is monitored as a function of the wavelength of the electronic excitation. (c) Doppler profile: the wavelength for the photofragment detection is scanned over the resonance wavelength for monitoring the photofragments.

transition and that for the photofragment detection is scanned over the resonance wavelength for monitoring the photofragments to obtain their Doppler profile. There are two variants of this measurement. If the photodissociation cross section is a slowly varying function of the electronic excitation wavelength, the same laser can be used for both the electronic excitation and the Doppler-profile measurement. Otherwise, the wavelength of the electronic excitation laser is held fixed and that for the Doppler profile scanned. The one-dimensional (1D) velocity distribution of the ensuing photofragments is extracted from their Doppler profile. To obtain 2D and 3D distributions, photofragment translational spectroscopy (PTS) or velocity-map imaging (VMI) are utilized. These and other monitoring techniques are presented in the next sections.

The experimental setup utilized for the VMP experiments represented by Figure 3.1 is schematically depicted in Figure 3.2 (an additional, more detailed drawing of an experimental setup is given in Figure 4.1, where VMP of HI is presented). In addition to the measurements described above, where the vibrational population of the initial electronic state is monitored indirectly by the action spectrum, this population is often monitored directly. This is done

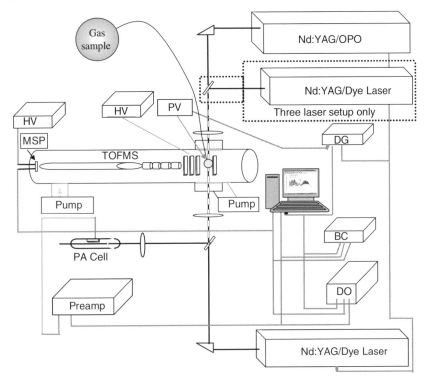

Figure 3.2 Schematic diagram of an experimental apparatus used in VMP experiments represented by Figure 3.1. Both a three-laser setup and a two-laser setup, where the same laser pulse is applied for both electronic excitation and detection of photofragments, are depicted. OPO=optical parametric oscillator; PA=photoacoustic; HV=high voltage; PV=pulsed valve; DG=delay generator; BC=boxcar averager; DO=digital oscilloscope; TOFMS=time-of-flight mass spectrometer; MSP=microsphere plate. The OPO produces IR photons for vibrational excitation, the other lasers UV photons for electronic excitation and REMPI. Reprinted from Ref. 28. Copyright 2005, with permission from Elsevier.

for wavelength calibration (the wavelengths for rovibrational transitions are well documented for many molecules), for assignment of the action spectrum and for elucidating the dynamics of VMP by comparing the spectra obtained in the two types of measurements. The methods applied for direct measurements of the vibrational populations in VMP studies are described below; one of them, photoacoustic (PA) spectroscopy, is alluded to in Figure 3.2. As shown in the figure, the residual IR beam of the Nd:YAG/optical parametric oscillator (OPO) laser, emerging from the time-of-flight mass spectrometer (TOFMS) after passing the interaction region, is used for the PA spectroscopy. We also note that the TOFMS, besides monitoring the REMPI products, is sometimes used for recording the parent REMPI (or one-step ionization) mass spectrum.

We now turn to describe in more detail experimental techniques that have been used in VMP studies, citing one example for each technique. We also mention some other techniques that may be applied in VMP studies.

3.1 Preparation and Detection of Vibrational States

3.1.1 Nonspecific Vibrational Excitation

In the early VMP-related studies, thermal vibrational excitation was utilized for understanding the temperature dependence of photoabsorption spectra and extracting the components of the spectra related to the contributions of the populated low-energy vibrational states. For example, measurements of the spectral absorption coefficient of OCS in the first absorption band around 226 nm were made at a series of temperatures from 195 to 404 K to determine whether the resulting changes in the shape of the spectrum can be attributed to contributions by specific vibrational levels of the lower electronic state.[1] Another method relies on the fact that electrical discharge sources often produce rotationally cold but vibrationally excited species. An example is the study of photodissociation of $SH(X\ ^2\Pi_{3/2},\ v''=2\text{--}7)$ and $SD(X\ ^2\Pi_{3/2},\ v''=3\text{--}7)$, produced by pulsed electrical discharge of D_2S, contaminated with H atoms in the gas-handling system.[2] In another nonspecific method, a beam of internally cold O_2^- anions was formed and then photodetached by a loosely focused pulsed laser beam resulting in formation of O_2 neutrals in a strongly inverted vibrational state distribution while remaining rotationally cold.[3]

There are additional methods for producing nonspecific vibrational excitation of the ground electronic state that, as far we are aware of, have not been applied thus far to VMP. (1) IR multiphoton excitation, where absorption of IR photons delivered by an intense laser enables deposition of many quanta of vibrational excitation with quite high efficiency but low specificity. (2) Electronic excitation followed by internal conversion to high-lying vibrational states of the ground electronic state. (3) FC pumping, where $v = 0$ in the ground electronic state is pumped to $v > 0$ in an excited electronic state and followed by emission to $v \gg 0$ in the ground electronic state. (4) Photodissociation of a polyatomic molecule that produces a vibrationally excited (smaller) molecular fragment. (5) Energy transfer from an electronically excited species to a molecule resulting in vibrational excitation of the latter. (6) Chemical reactions of the type $A + BC \rightarrow AB^\dagger + C$.

3.1.2 One-Photon, State-Selected Vibrational Excitation

One-step excitation is the most obvious means for preparing specific rovibrational states. IR photons with $v < \sim 3800\,cm^{-1}$ can efficiently excite all fundamental vibrations of molecules of interest in VMP provided the excitation is electric-dipole-allowed, *i.e.*, where the selection rules are $\Delta v = \pm 1$.[4,5] Excitation of overtones and combination bands is less efficient and decreases with increasing Δv (Section 1.2). Molecular gas lasers, *e.g.*, CO_2,[6] HBr,[7] and NH_3,[8] were used as a source for IR excitation of fundamental as well as low overtones of low-frequency modes in early VMP studies.

More recent studies have utilized various nonlinear optics techniques as well as tunable dye and solid-state lasers for exciting vibrational states in the IR, near-IR (NIR) and VIS regions. Thus, tunable IR radiation, obtained by Raman shifting the wavelength of an excimer-pumped dye laser (the dye laser output was focused in a high-pressure H_2 cell), was employed in the pioneering VMP studies of H_2O [9] for exciting its antisymmetric stretch around $3840\,cm^{-1}$. Difference-frequency mixing, in an appropriate nonlinear crystal, of a NIR laser beam (usually the 1.064-µm beam of a Nd:YAG laser) with a VIS beam, has been frequently used for obtaining an IR beam at wavelengths $> \sim 2\,µm$. This method was applied, for example, for obtaining $\sim 2.3\,µm$ for the investigation of VMP of HI (see Figure 4.1).[10]

For shorter wavelengths, needed for excitation of high overtones, both Nd:YAG-pumped tunable dye lasers and the idler beams of OPOs have been frequently used. An example of the former is excitation of the third stretching overtone ($v = 4$) of OH in HOD around 725 nm [11] and of the latter is excitation of C_2H_2 at $\sim 1\,µm$ in the region of three C–H stretch quanta.[12] Another example is the application of a Ti:sapphire laser, tunable at 870–880 nm, for exciting HF ($v = 3$).[13]

A single quantum-state-selected beam of oriented molecules can be obtained *via* hexapole state selection in symmetric-top molecules.[14] While traveling through the inhomogeneous electric field of the hexapole, molecules with a positive Stark effect experience a force towards the hexapole axis where the field strength vanishes. Thus, OCS ($v_2 = 0,1,2|J/M$) molecules were selected and focused in the photodissociation region by choosing a proper hexapole voltage.[15]

3.1.3 Two-Photon, State-Selected Vibrational Excitation

3.1.3.1 *Overtone–Overtone Double Resonance*

One-photon, selective excitation of high-overtone (or combination) states is difficult since absorption decreases with increasing vibrational state. Overtone–overtone double-resonance (or higher resonances) excitation schemes allow access to very high vibrational states and states of symmetry different from that possible by one-photon absorption. An example is the preparation of rovibrational states in the electronic ground state of H_2O at previously inaccessible energies.[16] This was achieved by first promoting H_2O in a particular rotational state to an intermediate level that contains four or five vibrational quanta in one of the OH stretches, then promoting a fraction of these pre-excited molecules to a higher rovibrational level containing between 8 and 12 OH-stretch quanta and between 0 and 1 quanta of OH bend. An illustration of this procedure is depicted in Figure 6.5.

3.1.3.2   *Preparation by Stimulated Raman Excitation and Detection by Coherent Anti-Stokes Raman Spectroscopy and Photoacoustic Raman Spectroscopy*

Stimulated Raman excitation (SRE) is a method by which a molecule can be excited to a fundamental or low-overtone vibration *via* interaction with two spatially and temporally overlapping laser beams at frequencies ω_p (pump) and ω_S (Stokes), with a difference frequency $(\omega_p - \omega_S)$ resonant with a Raman-active vibrational transition. The selection rules for Raman processes differ from those for direct IR, one-photon excitation. In particular, excitation of symmetric stretches or bends is allowed for the former rather than for the latter and the opposite is true for antisymmetric states.[4,5] The selection of rovi-brationally excited states is done by scanning the frequency ω_S of the Stokes laser. Rovibrational excitation by SRE can be monitored by coherent anti-Stokes Raman spectroscopy (CARS), a four-wave mixing process that "naturally" accompanies SRE. This occurs since ω_p can monitor the vibrationally excited state: the waves of frequencies ω_p, ω_S, ω_p are mixed in the sample to generate a new coherent wave at frequency $\omega_{CARS} = \omega_p - \omega_S + \omega_p$ (ω_{CARS} is an anti-Stokes beam). Although CARS may involve four waves of four different frequencies, the above scheme has been employed in VMP studies, *e.g.*, of HOD where the O–H ($\sim 3700\,\mathrm{cm}^{-1}$) and O–D ($\sim 2700\,\mathrm{cm}^{-1}$) stretches were excited by SRE.[17] An illustration of the SRE and CARS scheme, where two lasers (of frequencies ω_p and ω_S) are used is depicted in Figure 3.3.

There are several advantages in using SRE and CARS rather than direct IR excitation. First, as mentioned above, symmetric stretches can be excited, as was done, for example, in the VMP of H_2O (1,0,0).[18] Also, powerful lasers in the visible are more accessible than high-power mid-IR laser needed for exciting

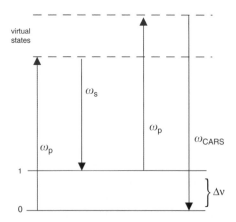

Figure 3.3   SRE and CARS schemes, where two lasers of frequencies ω_p and ω_S are used. Δv is the frequency of the vibrational state excited by the SRE process induced by the ω_p and ω_S beams represented by the two arrows in the left. ω_p also induces the CARS transition.

low-lying vibrational states. And last but not least, the coherent (laser) CARS signal can be monitored at large distances from the sample.

A related method for monitoring rovibrational states prepared by SRE is photoacoustic Raman spectroscopy (PARS). PARS saves the use of the optical components needed for monitoring the CARS signal and combines the advantages of SRE as an excitation method with the sensitivity of PA (described below) as a detection method. A comparison of CARS and PARS as a means for studying vibrational spectroscopy is presented in Ref. 19 and application of PARS to VMP (where the fundamental CH_3 stretch of CH_3CFCl_2 was pre-excited) in Ref. 20.

3.1.3.3   *Preparation by Stimulated Emission Pumping*

This method relies on FC pumping to an excited electronic state followed by stimulated dumping of this state to a high vibrational level of the ground electronic state.[21] The promotion of molecules in stimulated emission pumping (SEP) is by two photons of different wavelengths and the reachable vibrational states correspond to quite different skeletal motions of the molecule than those accessible in direct excitation. The pump laser acts first, followed, after some time delay, by the dump (or Stokes) laser. If the laser intensities are sufficiently high to saturate the transitions, 50% of the population of state $|1\rangle$ can be transferred to $|2\rangle$ of which 50% can be transferred to $|3\rangle$, *i.e.* the transfer efficiency from $|1\rangle$ to $|3\rangle$ can reach 25%. SEP has been used extensively as a spectroscopic method to provide new insights into IVR[22] and at least once in VMP studies, for pre-excitation of vibrational states in the VMP of CS_2.[23] An illustration of this procedure is depicted in Figure 5.3.

3.1.3.4   *Preparation by Stimulated Raman Adiabatic Passage*

Stimulated Raman adiabatic passage (STIRAP)[24–26] uses coherent, partially overlapping light pulses to produce highly efficient (up to 100%) population transfer between quantum states of an atom or a molecule. The procedure relies on an initial creation of coherence with subsequent adiabatic passage between the relevant states. In its simplest form the procedure involves a three-state, two-photon Raman process, in which an interaction with a pump pulse P links the initial state $|1\rangle$ with an intermediate state $|2\rangle$, which in turn interacts *via* a Stokes pulse S with the final state $|3\rangle$. Although STIRAP might resemble SEP, it is different. Unlike SEP, The Stokes laser acts first, followed, after a time delay that allows partial overlap, by the pump laser. Moreover, whereas for SEP coherence properties of the radiation are not relevant, coherence is essential for STIRAP. Most importantly, when STIRAP is applied the population of the intermediate state $|2\rangle$ is negligible, whereas for SEP it is significant. We note that when continuous-wave (CW) lasers are used in combination with molecular beams, spatially displaced, but partially overlapping, laser beams, propagating perpendicular to the molecular beam, achieve the partial temporal overlap. Since coherence is essential, single-mode CW lasers are used in this case. When pulsed lasers are used, the axes of the laser beams coincide and the pulses are delayed in time, and coherence (inferior to that of CW lasers) is achieved by transform-limited bandwidth of the pulses.

The method has been applied most successfully to atoms and diatomic molecules.[25,26] For excitation of polyatomics STIRAP has limitations, since in the case of increased density of states, particular transitions possess sub-stantially smaller line strength, requiring adequate laser power to achieve efficient transfer.[25,26] Still, a transfer efficiency of nearly 100% was demonstrated from the $(J,K) = (3,3)$ of the ground vibrational state (0,0,0) to the same rotational level of a high overtone (9,1,0) in SO_2.[25–27] STIRAP may be a useful means for VMP studies when other methods fail, but it has not been used thus far for this purpose.

3.1.4 Monitoring Vibrational Excitation

As mentioned above, vibrationally excited states are sensitively detected *via* the action spectrum, as well as by CARS and PARS. However, simpler, more direct measurements are often needed. Changes in the electronic absorption from the ground electronic state following vibrational excitation can be measured in order to monitor the vibrational population, as was done in NH_3 VMP-related studies.[6] However, PA spectroscopy is a much more sensitive method for monitoring vibrational excitation. The PA signal is produced as a result of energy transfer from the vibrational mode to translation, followed by local heating that induces pressure (sound) waves that are detected by a microphone. Indeed, PA spectroscopy has been commonly used in VMP studies, see, for example Ref. 28 and Figure 3.2 above. A special case of PA, PARS, was alluded to in Section 3.1.3.2.

3.2 Excitation and Detection of Electronic States

It was already noted that in order to start the electronic excitation from vibrationally pre-excited states rather than from the ground vibrational state, the wavelength of the photodissociating photon has to be at the leading edge of the absorption band of the molecule. Consequently, VMP, which is usually studied in the first (mostly UV) absorption band, is conducted using mid-UV and VIS wavelengths, the shortest wavelength being $\sim 193\,nm$ provided by the ArF excimer laser.[9] Longer wavelengths are mostly provided by the doubled output of a Nd:YAG-pumped dye laser. Frequently, the wavelength of the photodissociating laser is chosen to fit the $(2+1)$ REMPI transitions of the photofragments (Section 3.3.2), for example $\sim 243\,nm$ for detecting H atoms[12] and $\sim 235\,nm$ for Cl atoms.[20]

The electronic excitation is usually monitored indirectly *via* the detection of the photofragments (Section 3.3). From the information on the fragments (their internal-state distribution, translational energies and vector correlations) the quantum state of the excited electronic state from which the fragmentation took place can be inferred. When a UV action spectrum is employed, the excited electronic state is scanned and the dependence of the photofragment yield on specific rovibrational states in this electronic state is thus monitored.[12,28]

3.3   Detection and Characterization of Photofragments

3.3.1   Spontaneous Emission and Laser-Induced Fluorescence

If the photofragments are electronically excited and radiation to lower states is allowed, they may be detected *via* their emission, as was done in one of the VMP studies on NH_3 where the photodissociation was monitored *via* emission from the $NH_2(\tilde{A}\ ^2A_1)$ photofragment.[29] However, in most cases the photofragments are not electronically excited or only weakly radiate. Nevertheless, if a photofragment can be excited to a higher, strongly radiative electronic state, laser-induced fluorescence (LIF) is a sensitive detection method and has been frequently applied, for example in the VMP of H_2O, where the OH fragment was monitored at ~ 308 nm.[9]

3.3.2   Resonantly Enhanced Multiphoton Ionization

The detection of photofragments by resonantly enhanced multiphoton ionization (REMPI) is a highly sensitive method since it combines resonant excitation of a neutral species with subsequent ionization of this species and detection of ions. In its simplest form, which has been commonly applied in VMP studies, it involves *resonant excitation* by m photons and ionization, by n photons at the same wavelength, of the resonantly excited species (see Figures 1.1 and 3.1). It is then denoted as $(m+n)$ REMPI (in principle each of the $m+n$ photons could be of a different wavelength).

For $(2+1)$ REMPI (which is most often employed in VMP studies), the two-photon probability for a transition between an initial state i and an excited state e is given by[30]

$$S_{ie} = \sum_k \frac{\langle i|\boldsymbol{\mu}|k\rangle\ \langle k|\boldsymbol{\mu}|e\rangle}{E_k - E_i - h\nu} \tag{3.1}$$

where $\boldsymbol{\mu}$ is the electronic transition dipole moment operator and k denotes an intermediate, virtual state with energy E_k between i and e (half-way between them at resonance). As is obvious from eqn (3.1), when the transitions in the numerator are strong and the denominator $\to 0$, the $i \to e$ transition will be enhanced, consequently inducing selective ionization of the resonantly excited state.

The ions in the VMP experiments are usually monitored in a TOFMS, mostly of the Wiley–McLaren type.[31] The combination of REMPI with TOFMS provides high selectivity resulting from the resonant step and the mass-selective detection of the photofragments. It also enables monitoring the velocity distribution of specific photofragments.

These advantages of combining REMPI with TOFMS are represented, for example, in the VMP of $CHFCl_2$, where the two spin-orbit states of the two isotopes of the Cl atomic fragments were monitored.[32]

3.3.3 Doppler Profiles

The Doppler profiles of the photofragments are measured, primarily, in order to extract their kinetic-energy distribution. As described in Figure 3.1(c), for this measurement one wavelength (the REMPI wavelength in the figure) is scanned across the photofragment resonance, while the other wavelengths are held fixed. A 1D velocity distribution of the ensuing photofragments can thus be extracted from the profiles using the relation

$$\nu = \nu_0 \left(1 - \frac{v_z}{c}\right) \tag{3.2}$$

where ν is the laser frequency, v_z the projection of the fragment velocity on the laser-beam axis and ν_0 is the fragment resonance absorption frequency (at rest). Thus, in an ensemble of species the spread of absorption frequencies is related to their velocity distribution. When a Doppler profile exhibits a Gaussian line shape it indicates a Maxwellian speed distribution. In this case the distribution must be isotropic and the kinetic energy of the photofragment can be extracted from the profile. In addition, from the area ratios of the Doppler profiles of different species, the branching ratio for their production can be determined.

However, when the distribution is anisotropic, the shape of the profile is complex and from its analysis the involved dynamics of the VMP process can be analyzed. The elucidation of the vector correlations in VMP studies from the Doppler profiles was already mentioned (Section 2.1.4) and we note that in this respect the 1D profiles relay information on spatial anisotropy. An example is the evidence for the onset of three-body decay in photodissociation of vibrationally excited $CHFCl_2$.[32] We also note that when the same laser pulse is used for photodissociation and for monitoring the photofragments (steps (2) and (3) in Figure 3.1), the Doppler profile may be distorted due to the simultaneous scanning of the excited electronic state.[28]

3.3.4 High-*n* Rydberg Time-of-Flight

In the high-*n* Rydberg time-of-flight (HRTOF) technique[33] species in a pulsed molecular beam are excited to high-*n* Rydberg states. A small percentage of the excited species drift with their nascent velocities to a multichannel plate (MCP) detector, where they are detected as ions after being field ionized in front of the MCP.

Since the excited atoms are neutral, they are not affected by stray fields and space charge, which are major problems when detecting TOF of ions following laser ionization in the photodissociation region as in the REMPI procedure described above. This results in a much better translational energy resolution when ions are detected *via* HRTOF.

The technique was applied, for example, to the H atom photofragments in the VMP of C_2H_2.[34] The H atoms were excited to high-n Rydberg states by sequential absorption of 121.6-nm and 366-nm photons and then field ionized. From the center-of-mass translational-energy distributions the yields of $C_2H(\tilde{A})$ and $C_2H(\tilde{X})$ were obtained.

3.3.5 Photofragment Translational Spectroscopy

Photofragment translational spectroscopy (PTS) is a technique in which both the velocity and the angular distributions of ionized species in a molecular beam, mass selected in a mass spectrometer, are measured. This is done by monitoring the TOF over a series of angles obtained by rotating the electric vector of the polarized photolysis laser radiation[35] and/or rotating the molecular-beam sources or the detectors.[36] PTS was applied, for example, in the VMP of O_2.[3]

3.3.6 Velocity-Map Imaging

Velocity-map imaging (VMI) techniques (often referred to as velocity-map ion imaging) record 2D velocity projections of a given product and enable detailed study of any alignment and/or orientation of the products and of any correlations between the product $\mathbf{v}$ and $\mathbf{J}$ vectors.[37] These techniques use a charge coupled device (CCD) detector to collect the angle–velocity distribution of the product all at once. The ion optics are set for "velocity mapping", namely, each point on the detector corresponds to a specific 2D velocity coordinate. If the 3D velocity distribution is cylindrically symmetric, it is possible to construct it from the 2D projection that the detector measures.[37] VMI was employed, for example, in the VMP of SH and SD.[2]

The application of VMI to VMP has been increased during recent years. As we will see in the next chapters, VMI has been applied in VMP of diatomic, triatomic, tetratomic and larger molecules. We therefore present here some details about this technique, following closely the detailed description given in Ref. 37. When the parent molecules are photolyzed by a pulsed laser, the resulting photofragments are created in a localized volume, at a well-defined instant in time, and begin to expand in a series of concentric spheres the radii of which (at any particular instant in time) will be determined by the recoil velocities. These are known as Newton spheres. The VMI approach depicted in Figure 3.4 involves the following experimental steps: (a) creation of Newton spheres by photodissociation; (b) conversion of the photofragment spheres to ion spheres by laser ionization; (c) projection of the ion Newton spheres onto a 2D detector; and (d) recovery of the desired information from the 2D image,

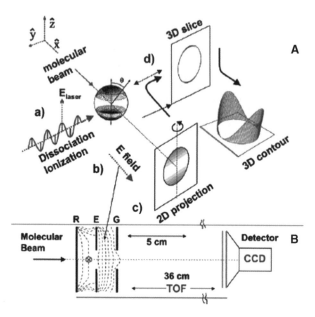

Figure 3.4 A: Schematics of VMI for measuring Newton spheres from photo-dissociation. B: Apparatus layout showing the electrostatic lens used for velocity-map imaging of photodissociation. TOF=time-of-flight; CCD= charge coupled device. See text for some further details and Ref. 37 for full details. Reproduced from Ref. 37 by permission of the PCCP Owner Societies.

either directly from the middle slice of the ion packet (experimental slicing), or *via* a mathematical transformation. The result of (d) should be equivalent to taking a thin slice through the middle of the Newton sphere formed in step (a).

References

1. J. R. Locker, J. B. Burkholder, E. J: Bair and H. A. Webster III, *J. Phys. Chem.*, 1983, **87**, 1864.
2. L. M. C. Janssen, M. P. J. van der Loo, G. C. Groenenboom, S-M. Wu, D. Č. Radenović, A. J. A. van Roij, I. A. Garcia and D. H. Parker, *J. Chem. Phys.*, 2007, **126**, 094304.
3. D. J. Leahy, D. L. Osborn, D. R. Cyr and D. M. Neumark, *J. Chem. Phys.*, 1995, **103**, 2495.
4. G. Herzberg, *Molecular Spectra and Molecular Structure I. Spectra of Diatomic Molecules*, Van Nostrand, Princeton, 1950.
5. G. Herzberg, *Molecular Spectra and Molecular Structure II. Infrared and Raman Spectra of Polyatomic Molecules*, Van Nostrand, Princeton, 1945.
6. R. V. Ambartzumian, V. S. Letokhov, G. N. Makarov and A. A. Puretzkiy, *Chem. Phys. Lett.*, 1972, **16**, 252.

7. P. F. Zittel and D. D. Little, *J. Chem. Phys.*, 1979, **71**, 713.

8. P. F. Zittel and D. E. Masturzo, *J. Chem. Phys.*, 1986, **85**, 4362.

9. D. Häusler, P. Andresen and R. Schinke, *J. Chem. Phys.*, 1987, **87**, 3949.

10. J. P. Camden, H. A. Bechtel, D. J. A. Brown, A. E. Pomerantz, R. N. Zare and R. J. Leroy, *J. Phys. Chem. A*, 2004, **108**, 7806.

11. R. L. Vander Wal, J. L Scott, F. F. Crim, K. Weide and R. Schinke, *J. Chem. Phys.*, 1991, **94**, 3548.

12. Y. Ganot, A. Golan, X. Sheng, S. Rosenwaks and I. Bar, *Phys. Chem. Chem. Phys.*, 2003, **5**, 5399.

13. J. Zhang, C. W. Riehn, M. Dulligan and C. Wittig, *J. Chem. Phys.*, 1996, **104**, 7027.

14. M. H. M. Janssen, J. W. G. Mastenbroek and S. Stolte, *J. Phys. Chem. A*, 1997, **101**, 7605.

15. M. L. Lipciuc and M. H. M. Janssen, *J. Chem. Phys.*, 2007, **126**, 194318.

16. P. Maksyutenko, J. S. Muenter, N. F. Zobov, S. V. Shirin, O. L. Polyansky, T. R. Rizzo and O. V. Boyarkin, *J. Chem. Phys.*, 2007, **126**, 241101.

17. I. Bar, Y. Cohen, D. David, T. Arusi-Parpar, S. Rosenwaks and J. J. Valentini, *J. Chem. Phys.*, 1991, **95**, 3341.

18. D. David, A. Strugano, I. Bar and S. Rosenwaks, *J. Chem. Phys.*, 1993, **98**, 409.

19. A. Melchior, I. Bar and S. Rosenwaks, *J. Phys. Chem. A*, 1998, **102**, 7273.

20. A. Melchior, X. Chen, I. Bar and S. Rosenwaks, *Chem. Phys. Lett.*, 1999, **315**, 421.

21. C. H. Hamilton, J. L. Kinsey and R. W. Field, *Annu. Rev. Phys. Chem.*, 1986, **37**, 493.

22. D. J. Nesbitt and R. W. Field, *J. Phys. Chem.*, 1996, **100**, 12735.

23. H. T. Liou, Y. C. Chang and Z. Liou, *J. Phys. Chem. A*, 2006, **108**, 4610.

24. J. Oreg, F. T. Hioe and J. H. Eberly, *Phys. Rev. A*, 1984, **29**, 690.

25. K. Bergmann, H. Theuer and B. W. Shore, *Rev. Mod. Phys.*, 1998, **70**, 1003.

26. N. V. Vitanov, M. Fleischhauer, B. W. Shore and K. Bergmann, *Adv. Atom. Molec. Opt. Phys.*, 2001, **46**, 55.

27. T. Halfmann and K. Bergmann, *J. Chem. Phys.*, 1996, **104**, 7068.

28. A. Portnov, Y. Ganot, S. Rosenwaks and I. Bar, *J. Mol. Struct.*, 2005, **744–747**, 107.

29. A. Bach, J. M. Hutchison, R. J. Holiday and F. F. Crim, *J. Chem. Phys.*, 2002, **116**, 4955.

30. J. Pfab, in *Spectroscopy in Environmental Science*, ed. R. J. H. Clark and R. E. Hester, Wiley, New York, 1995, p 149.

31. W. C. Wiley and I. H. McLaren, *Rev. Sci. Instrum.*, 1955, **26**, 1150.

32. X. Chen, R. Marom, S. Rosenwaks, I. Bar, T. Einfeld, C. Maul and K-H. Gericke, *J. Chem. Phys.*, 2001, **114**, 9033.

33. L. Schneider, W. Meier, K. H. Welge, M. N. R. Ashfold and C. Western, *J. Chem. Phys.*, 1990, **92**, 7027.

34. J. Zhang, C. W. Riehn, M. Dulligan and C. Wittig, *J. Chem. Phys.*, 1995, **103**, 6815.

35. S. J. Riley and K. R. Wilson, *Faraday Discuss. Chem. Soc.*, 1972, **53**, 132.
36. A. M. Wodtke and Y. T. Lee, in *Molecular Photodissociation Dynamics*, ed.
     M. N. R. Ashfold and J. E. Baggott, Royal Society of Chemistry, London,
     1987, p 31.
37. M. N. R. Ashfold, N. H. Nahler, A. J. Orr-Ewing, O. P. J. Vieuxmaire,
     R. L. Toomes, T. N. Kitsopoulos, I. A. Garcia, D. A. Chestakov, S-M. Wu
     and D. H. Parker, *Phys. Chem. Chem. Phys.*, 2006, **8**, 26.

CHAPTER 4
VMP of Diatomic Molecules and Radicals

VMP of diatomic molecules and radicals is discussed here only briefly since mode-selective or bond-selective dissociation, which has been the main reason for the recent interest in VMP, is not relevant for these molecules. The aim of VMP studies of diatomic molecules, starting in the 1970s, has been to investigate the effect of rovibrational parent excitation on the photodissociation dynamics, *e.g.*, enhancement of the dissociation cross section, polarization dependence of product angular distribution and branching ratios between electronic states of the products. However, already in the 1920s, in the seminal papers of Franck[1] and Condon,[2,3] the contribution of vibrational (thermal) population to the continuous absorption (leading to dissociation) in homonuclear and heteronuclear diatomic molecules was noted. This was followed by detailed analysis of the temperature-dependent continuous absorption of Cl_2,[4–7] Br_2[8] and Na_2.[8a] For Na_2 the analysis considered in detail the variation of the electronic dipole moment as a function of the internuclear separation, following previous studies on photodissociation of Na_2.[8b–8e] Good agreement between experiment and theory was found for the relative variation of the absorption cross section but not for the absolute values.[8a] The discrepancy was attributed to inaccuracies in the measurements and in the PES used.

An illuminating, early application of the FC principle is presented in Refs. 8f and 8g for the photodissociation of vibrationally (thermally) populated NaI. The dependence of the recoil velocity of the Na fragment on the vibrational level of the NaI parent was calculated. An oversimplified form of the FC principle was applied, which assumes that electronic transitions occur only from the midpoint of the classical vibration when $v = 0$, and only from the turning points when $v > 0$. Nevertheless, this is, in a sense, a state-to-state VMP (computational) study.

Vibrationally Mediated Photodissociation
By Salman (Zamik) Rosenwaks
© Salman (Zamik) Rosenwaks 2009
Published by the Royal Society of Chemistry, www.rsc.org

Preparation of *specific* vibrational levels was applied in VMP studies on several diatomic molecules: all four hydrogen halides were studied experimentally (HF,[9] HCl,[10–11a,11b] HBr,[12,12a] and HI[13]) and theoretically (HF[14,14a,14b] and DF,[14,14a] HCl,[11a,11b,11c,15,15a] HBr,[16,16a] and HI[17–20b] and DI.[17,20a]). In addition, the VMP of H_2,[15a] IBr,[18a,18b] OH[20c,20d,20e] and OD[20e,20f] was theoretically, and of OH,[20e] OD,[20e,20f] SH,[21,22] SD[22] and O_2[23,24] experimentally studied. Some of the experimental studies and their analysis are briefly summarized below. Unless otherwise stated, it is assumed that the two atomic photofragments were produced in their ground electronic states.

4.1   HF

Spin-orbit branching ratios of the F photofragments and infrared alignment of the parent were studied in VMP of HF.[9] Single rotational levels of HF($X\ ^1\Sigma^+$, $v = 3$) molecules were prepared by using overtone excitation by an IR laser and the molecules were then photodissociated by UV radiation at 193 nm. The HRTOF technique was applied to the H atom product to obtain center-of-mass translational-energy distributions and provided the spin-orbit state distribution of the fluorine fragment. The hyperfine coupling of HF in $v = 3$ was examined, testing the vibrational dependence of the coupling constants. Contributions from higher-order terms in the vibration were suggested. The dependence of the H atom angular distribution on IR laser polarization indicated a pronounced alignment, whose time dependence was examined by varying the delay between the IR and UV lasers. Thus, in addition to HF $v = 3$ alignment parameters measured in this experiment, it also provided the first experimental demonstration of alignment depolarization, due to hyperfine interaction, in this system.

4.2   HCl

The apparently first published two-step (using IR-UV lasers) VMP study was carried out on HCl($X\ ^1\Sigma^+$).[10,10a] An IR laser excited the $v = 3$ state of HCl, while a UV laser photodissociated only the vibrationally excited molecule. The Cl photofragments reacted with NO to produce NOCl.

In a much later, more elaborate, study the effect of internal parent excitation on the UV-photodissociation dynamics of HCl($X\ ^1\Sigma$) molecules was investigated in jet-cooled H^{35}Cl molecules within a TOFMS.[11] The HCl was excited to a particular rovibrational level of the ground electronic state using tunable IR-laser radiation. Wavelengths required for the first and second overtone bands of HCl were directly provided by the idler output of an OPO. The fundamental vibrational band wavelengths were obtained by difference-frequency mixing of visible laser beams. A laser beam of UV light ~ 235 nm subsequently photolyzed the prepared HCl (v, J) molecules and state-selectively ionized the Cl(^2P_J) photofragments *via* (2 + 1) REMPI. The production of Cl$^+$

ions was monitored using a TOFMS as the UV wavelength was scanned over the full Cl atom spectral feature. The following quantum states were prepared by infrared absorption: $v=1$, $J=0$ and $J=5$; $v=2$, $J=0$ and $J=11$; $v=3$, $J=0$ and $J=7$. The results were presented as the fraction of total chlorine yield formed in the spin-orbit excited state, $Cl(^2P_{1/2})$. Although there is not a strong or clear-cut v dependence on the branching fraction in this excitation region for HCl, calculations[11] suggest that there should be very pronounced effects on the branching fraction at higher excitation energies.

Preparation of highly polarized nuclei and optical control of atomic orbital alignment of $^{35}Cl(^2P_{3/2})$ atoms has also been demonstrated in VMP of HCl.[11a,11b] In the first experiment[11a] $H^{35}Cl(v=2, J=1, M=1)$ state was prepared with a 1.7-μm laser pulse and then dissociated with a delayed 235-nm laser pulse to produce ^{35}Cl atoms. Time-dependent polarizations of both $H^{35}Cl(v=2, J=1)$ molecules and $^{35}Cl(^2P_{3/2})$ atoms, which varied due to hyperfine quantum beating, were measured. The ^{35}Cl nuclear spin was highly polarized at a pump-probe delay of 145 ns. In the second experiment[11b] $H^{35}Cl(v=2, J=1, M=0)$ was prepared. These rotationally aligned $J=1$ molecules were then selectively photodissociated with a linearly polarized laser pulse at 220 nm after a time delay and the velocity-dependent alignment of the $^{35}Cl(^2P_{3/2})$ photofragments measured (the $^{35}Cl(^2P_{3/2})$ atoms are aligned by two mechanisms: (1) the time-dependent transfer of rotational polarization of the $H^{35}Cl(v=2, J=1, M=0)$ molecule to the $^{35}Cl(^2P_{3/2})$ nuclear spin contributes to the total $^{35}Cl(^2P_{3/2})$ photofragment atomic polarization and (2) the alignment of the $^{35}Cl(^2P_{3/2})$ electronic polarization resulting from the photoexcitation and dissociation process). The total alignment of the $^{35}Cl(^2P_{3/2})$ photofragments was found to vary as a function of time delay between the excitation and the photolysis laser pulses, in agreement with theoretical predictions. It was shown that the alignment of the ground-state $^{35}Cl(^2P_{3/2})$ atoms, with respect to the photodissociation recoil direction, can be controlled optically. It was noted that potential applications include the study of alignment-dependent collision effects.

4.3 HBr

In a pioneering, quantitative work on VMP of a diatomic molecule, the relative absorption cross sections of $HBr(X\ ^1\Sigma^+, v=1)$ and $(v=0)$ at 258.9 nm were compared.[12] The $v=1$ state was prepared *via* IR excitation by an HBr pin-discharge laser. The absorption cross section for HBr $(v=1)$ was found to be $(5.5\pm1.1)\times10^{-20}$ cm^2/molecule, 34 ± 7 times larger than that for $(v=0)$ at this wavelength. A theoretical estimate of the ratio of cross sections, based on applying the reflection principle[25] to the UV absorption spectrum of HBr, gave poor agreement with the experimental result. The importance of this experiment stems from the fact that it directly and quantitatively measured, for the first time, the enhancement of the photodissociation cross section following vibrational excitation to a specific vibrational state.

4.4 HI

The photolysis of HI($X\ {}^1\Sigma^+$, $v = 2$, $J = 0$) $\rightarrow$ H + I($^2P_{1/2}$)/I($^2P_{3/2}$) was studied over the wavelength range 297–350 nm.[13] Vibrationally excited HI was prepared using direct absorption of an IR beam, obtained by difference-frequency mixing of visible laser beams amplified in an optical parametric amplifier (OPA), and the H-atom photofragments were probed *via* REMPI coupled with core-extraction TOFMS. Schematics of the experimental setup used for this detailed study are depicted in Figure 4.1.

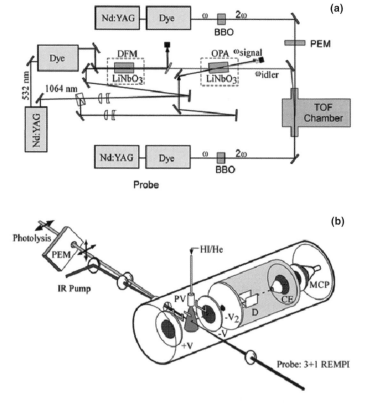

Figure 4.1 Schematics of the experimental setup for the VMP studies of HI. (a) Block diagram of the laser systems used to generate the IR, photolysis, and probe light. The IR is generated by difference-frequency mixing, DFM, and optical parametric amplification, OPA. The photolysis and probe light is generated by frequency doubling in BBO the output of a dye laser. The photoelastic modulator, PEM, is used for rotating the polarization of the laser beam. TOF=time-of-flight. (b) Schematic of a Wiley–McLaren TOF mass spectrometer: D=deflection plates, CE=core extractor, MCP=multichannel plates, and PV=pulsed valve. Reprinted with permission from Ref. 13. Copyright 2004 American Chemical Society. 2004, **108**, 7806.

Both the spin-orbit branching ratios of the I atoms and the photofragment angular distribution were measured. The branching-ratio measurements disagreed with predictions obtained from *ab initio* calculations and from the results of an empirical analysis based on experimental values of the $HI/DI(v=0)$ absorption cross sections and branching ratios. This data was combined with all existing absorption coefficient and branching fraction data for HI/DI in a global analysis that provided a new empirical determination of the final-state potential curves and transition moment functions for the four excited electronic states contributing to the A-band UV absorption continuum of HI. The analysis yielded two models for the radial dependence of the excited-state potential-energy curves and transition dipole moment functions. The existing data could not differentiate these models, but the study identified a range of experiments that would do so. It was demonstrated, once again, that photodissociation of vibrationally excited molecules is a sensitive probe of the repulsive excited electronic states encountered in bound–free transitions.

However, from a more recent analysis[20b] it was concluded that a simple and theoretically meaningful correction, namely, a small shift of the A-band triplet states of HI to higher energies, essentially removes the discrepancy between theory and experiment and the HI photodissociation branching ratios agree quite well with the measured data.

4.5 OH, OD, SH and SD Radicals

Vibrationally excited OH/OD (Refs. 20e/20e,f), produced in a pulsed electric discharge in a supersonic beam of H_2O/D_2O (seeded in Ar), were state-selected and focused by a hexapole and then photodissociated by a single laser tuned to various H/D or O atom (2+1) REMPI wavelengths between 243 nm and 200 nm. The angle velocity distributions of the resulting O^+ and D^+ photofragment ions were recorded using VMI. Photodissociation to the $O(^3P_J) + H/D(^2S)$ limit was shown to take place by one-photon excitation to the repulsive 1 $^2\Sigma^-$ state. The experimental data shows that vibrationally excited OH/OD (up to $v''=3/5$), formed in the discharge, are dissociated (a vibrational temperature of 2000 K was estimated for the beam source). It is noteworthy that while the photodissociation of OH/OD is energetically allowed at $\lambda < 284$ nm, FC overlap restricts absorption from OH/OD (X, $v''=0$) to $\lambda < 180$ nm. However, when the X state is populated in higher vibrational levels, $\lambda > 180$ nm radiation can excite the radicals into the repulsive 1 $^2\Sigma^-$ state, leading to the observed H/D and O atom product signals.

UV photodissociation of SH(X $^2\Pi_{3/2}$, $v''=0$–2), generated by 193-nm photolysis of H_2S, was studied at 216–232 nm using HRTOF to probe the H atom translational distribution and infer the spin-orbit branching fractions of the $S(^3P_2)$ product.[21] Photodissociation of SH(X $^2\Pi_{3/2}$, $v''=2$–7) and SD(X $^2\Pi_{3/2}$, $v''=3$–7), produced by pulsed electrical discharge of D_2S, contaminated with H atoms in the gas-handling system, was studied at 288 and 291 nm using VMI to

14. A. Brown and G. G. Balint-Kurti, *J. Chem. Phys.*, 2000, **113**, 1879.

14(a). G. G. Balint-Kurti, A. J. Orr-Ewing, J. A. Beswick, A. Brown and O. S. Vasyutinskii, *J. Chem. Phys.*, 2002, **116**, 10760.

14(b). L. Rubio-Lago, D. Sofikitis, A. Koubenakis and T. P. Rakitzis, *Phys. Rev. A*, 2006, **74**, 042503.

15. S. C. Givertz and G. G. Balint-Kurti, *J. Chem. Soc., Faraday Trans. 2*, 1986, **82**, 1231.

15(a). T. P. Rakitzis, *Phys. Rev. Lett.*, 2005, **94**, 083005.

16. B. Pouilly and M. Monnerville, *Chem. Phys.*, 1998, **238**, 437.

16(a). W. C. Hwang, R. F. Kamada and C. C. Badcock, *Int. J. Chem. Kin.*, 1977, **9**, 969.

17. M. Shapiro, in *Isotope Effects in Gas-Phase Chemistry*, ed. J.A. Kaye, ACS Symposium Series 502, ACS, Washington D.C., 1992, p. 264.

17(a). C. Kalyanaraman and N. Sathyamurthy, *Chem. Phys. Lett.*, 1993, **209**, 52.

18. P. Gross, A. K. Gupta, D. B. Bairagi and M. K. Mishra, *J. Chem. Phys.*, 1996, **104**, 7045.

18(a). K. Vandana and M. K. Mishra, *J. Chem. Phys.*, 1999, **110**, 5140.

18(b). K. Vandana and M. K. Mishra, *J. Chem. Phys.*, 2000, **113**, 2336.

19. A. B. Alekseyev, H. P. Liebermann, D. B. Kokh and R. J. Buenker, *J. Chem. Phys.*, 2000, **113**, 6174.

20. H. Fujisaki, Y. Teranishi and H. Nakamura, *J. Theor. Comput. Chem.*, 2002, **1**, 245.

20(a). R. J. Le Roy and G. T. Kraemer, *J. Chem. Phys.*, 2002, **117**, 9353.

20(b). A. B. Alekseyev, D. B. Kokh and R. J. Buenker, *J. Phys. Chem. A*, 2005, **109**, 3094.

20(c). S. Lee, *J. Chem. Phys.*, 1996, **104**, 1912.

20(d). S. Lee, *J. Chem. Phys.*, 1997, **107**, 1388.

20(e). D. Č. Radenović, A. J. A. van Roji, S.-M. Wu, J. J. Ter Meulen, D. H. Parker, M. P. J. van der Loo, L. M. C. Janssen and G. C. Groenenboom, *Mol. Phys.*, 2008, **106**, 557.

20(f). D. Č. Radenović, A. J. A. van Roij, D. A. Chestakov, A. T. J. B. Eppink, J. J. Ter Meulen, D. H. Parker, M. P. J. van der Loo, G. C. Groenenboom, M. E. Greenslade and M. I. Lester, *J. Chem. Phys.*, 2003, **119**, 9341.

21. W. Zhou, Y. Yuan, S. Chen and J. Zhang, *J. Chem. Phys.*, 2005, **123**, 054330.

22. L. M. C. Janssen, M. P. J. van der Loo, G. C. Groenenboom, S. -M. Wu, D. Č. Radenović, A. J. A. van Roij, I. A. Garcia and D. H. Parker, *J. Chem. Phys.*, 2007, **126**, 094304.

23. D. J. Leahy, D. L. Osborn, D. R. Cyr and D. M. Neumark, *J. Chem. Phys.*, 1995, **103**, 2495.

24. R. Toumi, B. J. Kerridge and J. A. Pyle, *Nature*, 1991, **351**, 217.

25. R. Schinke, *Photodissociation Dynamics*, Cambridge University Press, Cambridge, 1993.

CHAPTER 5
VMP of Triatomic Molecules Excluding Water

Triatomic molecules are the simplest polyatomic where mode-selective or bond-selective dissociation can be demonstrated. Moreover, they are small enough to allow *ab initio* calculations of PESs and photodynamics although they possess several vibrational degrees of freedom, like stretches and bends, which play a principal role also in larger molecules. VMP of several triatomic molecules has been studied for many years (O_3,[1–38b] OCS,[14,39–61] OCSe,[62] N_2O,[46,63–72] CS_2,[46,73,74] ICN,[74a,b] DCN[75] and HCN,[75,75a] HOCl[75a,76] and DOCl,[76] and HOBr[77]) both experimentally and theoretically. However, it was the extensive theoretical and experimental investigations of the water isotopologues, in particular H_2O and HOD, that opened a new era of detailed studies of state-to-state photodissociation out of specific rovibrationally excited states of polyatomic molecules. The studies of the water isotopologues are discussed in detail in Chapter 6, following the present chapter that consists of short sections on the other triatomic molecules, except O_3 and OCS, which are discussed in more detail. O_3 due to the role of VMP in its solar photolysis and OCS since in this case vibrational-state specificity in VMP was studied quite extensively.

5.1 VMP of O_3

5.1.1 Spectral Features of O_3 Relevant to VMP

Ozone, O_3, is a bent molecule with a bond angle of 116.8° (Ref. 77a). The fundamental vibrations are $v_1 = 1103 \, cm^{-1}$ and $v_3 = 1042 \, cm^{-1}$ for the symmetric and antisymmetric stretch, respectively, and $v_2 = 701 \, cm^{-1}$ for the bending.[16] Early VMP studies of ozone were facilitated by the availability of CO_2 lasers at convenient frequencies for excitation of the v_3 level.

Vibrationally Mediated Photodissociation
By Salman (Zamik) Rosenwaks
© Salman (Zamik) Rosenwaks 2009
Published by the Royal Society of Chemistry, www.rsc.org

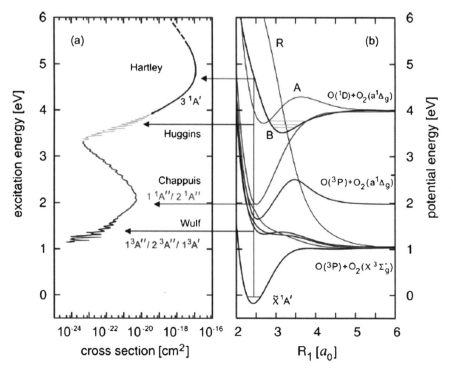

Figure 5.1   The photophysics of ozone for excitation energies up to 6 eV. (a) The measured absorption cross section (in cm^2; logarithmic scale) of ozone as a function of the excitation energy. (b) One-dimensional cuts through the PESs relevant for the photodissociation of ozone. R_1 is one of the O–O bond lengths; the other one is fixed at $R_2 = 2.43$ a_0 and the bond angle is $\alpha = 117°$. $E = 0$ corresponds to O$_3$($\tilde{X}$) in the ground vibrational state (zero-point energy). A, B, and R indicate the three (diabatic) $^1A'$ states relevant for the Hartley and Huggins bands. The horizontal arrows illustrate the electronic assignments of the absorption bands. Shading in (a) and (b) stresses the relation between the absorption bands and the underlying electronic states. Reproduced from Ref. 38a by permission of the PCCP Owner Societies.

As an introduction to VMP of ozone, its measured absorption cross section as a function of the excitation energy and one-dimensional cuts through the PESs relevant for the photodissociation of ozone are depicted in Figure 5.1 (Ref. 38a). This reference reviews recent theoretical studies of the photodissociation of ozone in the wavelength region from 200 nm to 1100 nm. It also cites many references on hot bands, *i.e.*, bands that originate from excited vibrational states in the ground electronic state and obviously are relevant to VMP of ozone. The measured absorption cross sections of ozone shown in Figure 5.1(a) are for the energy range from about 1 eV to 6 eV. Four distinct absorption bands can be identified: Wulf, Chappuis, Huggins, and Hartley. The UV radiation is absorbed by the strong Huggins–Hartley band system that is our main interest in discussing the VMP of ozone. All four bands exhibit more or less pronounced diffuse vibronic structures that reflect different details of the intramolecular vibrational dynamics in the particular excited electronic states and the nonadiabatic couplings between different states.

One of the main reasons for the interest in ozone photolysis is the production of electronically excited oxygen atom, $O(^1D)$, in the spin-allowed dissociation of O_3 molecules excited to the $1\ ^1B_2$ state, the so-called B state, at $\lambda \leq 310$ nm ($E \geq \sim 4$ eV, see Figure 5.1):[38a]

$$O_3(\tilde{X}\ ^1A') + h\nu \xrightarrow{\lambda \leq 310\,nm} O_3(1\ ^1B_2) \rightarrow O_2(a\ ^1\Delta_g) + O(^1D_2) \qquad (5.1)$$

$O(^1D)$ is an important species in the chemistry of the lower atmosphere.[27] The small fraction of $O(^1D)$, produced in the atmosphere by solar photolysis, that survives quenching to the ground state, $O(^3P)$, reacts with water and small hydrocarbons and turns out to be the major source of OH that, in turn, initiates the oxidation of pollutants, thereby cleansing the atmosphere. Knowledge of how $O(^1D)$ is formed in the atmosphere is therefore critical in understanding the creation of OH.

5.1.2   VMP of O_3 as a Source of $O(^1D)$

The importance of VMP of ozone stems from the fact that the longer-wavelength "tail" ($\lambda > 310$ nm) in its solar photolysis produces $O(^1D)$ *via* excitation of vibrationally excited ozone (it should be noted that also very weakly populated spin-forbidden channels producing $O(^1D)$ have been observed experimentally in the wavelength range corresponding to the Huggins band[27]). The longer wavelengths are important because stratospheric ozone screens most of the shortwave UV (which produces $O(^1D)$ from the vibrational ground state of ozone) from the lower atmosphere. Moreover, despite significantly lower populations in the vibrationally excited levels, greatly increased FC factors with excitation of the stretching modes result in relatively large intensities of hot bands. Thus, the observed $O(^1D)$ quantum yield from VMP of ozone is comparable to that observed for photodissociation of the ground-state molecules *via* (5.1).[27]

It is thus not surprising that numerous investigations of ozone photolysis were carried out, many of them alluding to vibrationally excited ozone.[1–38b] The effect of vibrational excitation, either thermal[1,3,5,9,12–38] or due to ground state O plus O_2 recombination,[2,4,11,16] on the shape of the absorption bands was examined.[1–4,9,11–38a] Moreover, the influence of vibrational excitation on the yield of the photofragments and on their kinetic energy was studied in detail using REMPI and LIF. The yields of, mostly, $O(^1D_2)$[5,17,20–28,30,31,37] and $O_2(a^1\Delta_g)$[17,18,19,26,38] but also $O(^3P_j)$[20,28,29,31,32] (and more specifically $O(^3P_0)$[29,32] and $O(^3P_2)$[31]) and $O_2(b^1\Sigma_g^+)$[29,32] and their kinetic energy[18,19,21,24,26,28–31,36–38] was monitored. In addition, various methods were used to calculate the effect of vibrational excitation on the shapes of the spectra and to compare them to experimental results.[1–4,9,11–38b] The experimental and theoretical investigations clearly confirmed that the photolysis of vibrationally excited ozone is efficient due to the high FC factors for the electronic excitation step and is a major source of $O(^1D)$ in the atmosphere.

For further understanding of VMP in ozone, an IR-UV two-laser photodissociation technique was used to measure the effect of ozone vibrational excitation on the cross section for photodissociative production of $O(^1D)$ at several UV wavelengths.[5–8,10] In these IR-UV double-resonance experiments a CO_2 laser, operating at $\sim 9.5\,\mu m$, populates the antisymmetric stretching mode v_3, (001), at $1042\,cm^{-1}$, which rapidly equilibrate with the symmetric stretching mode v_1, (100) at $1103\,cm^{-1}$. Below the ~ 310-nm energy threshold for the process given by eqn (5.1), vibrational excitation of O_3 was found to increase the cross section for production of $O(^1D)$ by nearly two orders of magnitude at 300 K.[5] Substantial increases in both the O_3 photoabsorption cross section and the $O(^1D)$ quantum yield were observed.[5] Moreover, although mode dependence of the VMP was not directly measured, it was concluded from a study of the photofragment excitation spectrum for $O(^1D)$ from the photodissociation of jet-cooled ozone in the wavelength range 305–329 nm, that the active mode for the hot-band excitation was the v_3 mode in the ground electronic state.[22] This conclusion was partially supported by a study of the dynamics of ozone photolysis at room temperature in the wavelength region 301–311 nm, where $O_2(^1\Delta_g)$ was probed.[38] It was estimated from this study that in the photodissociation of vibrationally excited ozone beyond the photodissociation threshold wavelengths of 310 nm, $\sim 1/3$ of the photofragments are from excitation in the v_3 mode.[38] However, a different conclusion has been presented following an analysis of the longer-wavelength tail of the ultraviolet absorption spectrum of O_3.[32] The structure of the absorption bands was assigned to vibrational progressions in the v_1 mode and the bending mode v_2, (010), at $701\,cm^{-1}$ but not the v_3 mode. This apparent disagreement points to the need for state-to-state VMP experiments to resolve such issues, as exemplified in studies of other molecules described in subsequent sections.

It is noteworthy in this context that from the theoretical analysis of Ref. 38a it seems that excitation of the v_3 antisymmetric stretch fundamental is more efficient than excitation of the v_1 symmetric stretch fundamental in increasing the overlap with the wavefunction in the upper state. We also note that the effect of excitation of the (1,0,0), (0,0,1), (0,1,0) and (0,2,0) vibrational modes on the photodissociation of ozone in the Hartley continuum has been recently theoretically studied and reported in Ref. 36 and corrected in 38b.

5.1.3 Vector Correlations in the VMP of O_3

Additional insight into VMP of ozone was obtained from studies of vector correlations.[29,37,38] The translational anisotropy (the β parameter),[29,37,38] electronic angular momentum polarization of $O(^1D_2)$[37] and the rotational angular momentum orientation of the $O_2(^1\Delta_g, v=0, J=16–20)$[38] obtained from photolysis of vibrationally excited ozone were studied. It was found[37] that for photolysis wavelengths greater than 310 nm, $O(^1D_2)$ is formed from the dissociation of internally excited ozone and the corresponding β parameters are markedly lower than for atomic fragments produced with the same speed from the photolysis of ground-state ozone. The general trend observed for the β parameters of the atomic fragment was also observed for the molecular photofragments.[38] The lower value of β for the dissociation of vibrationally excited ozone may be a consequence of a contribution to the dissociation from an excitation whose transition dipole moment possesses different symmetry from that for excitation from the vibrational ground state.[37] The angular-momentum polarization of $O(^1D_2)$ produced in VMP of ozone increases, which is consistent with the increasing contribution from the photolysis of internally excited ozone as the dissociation wavelength lengthens.[37] As for the rotational orientation moments, they were found to be negative for photolysis at 301 nm, indicating that a bond-opening mechanism provides the key torque for the departing O_2 fragment, and to become positive for photolysis beyond the dissociation threshold, as the increasing impulsive dissociation begins to dominate the nature of the rotation of the departing molecular fragment.[38]

5.2 VMP of OCS

5.2.1 Spectral Features of OCS Relevant to VMP

The fundamental vibrations of OCS are $v_1 = 2062\,\text{cm}^{-1}$ and $v_3 = 859\,\text{cm}^{-1}$ for the C–O and C–S stretch, respectively, and $v_2 = 520\,\text{cm}^{-1}$ for the bending.[77b] As in ozone, early VMP studies were facilitated by the availability of CO_2 lasers at convenient frequencies, which were suitable in the present case for excitation of the $2v_2$ level.

What makes the OCS molecule an interesting system for VMP studies is that its electronic ground state is linear, whereas its dissociation takes place through excitation to bent electronic states. For the bent molecule the symmetry of the linear OCS, $C_{\infty v}$, is lowered to C_s. The $^1\Sigma^+$ ground state is of A' symmetry, and the $1\,^1\Delta$ excited state splits into the $2\,^1A'$ and $2\,^1A''$ states (see Figure 5.2). The nearby (for $\theta = 0°$) $1\,^1\Sigma^-$ state correlates with the $1\,^1A''$ state. Electronic transitions originally forbidden in the linear geometry from the ground state are induced by bending distortion. To put it differently, due to the interaction between the motion of the electrons and the vibrational motion of the nuclei, known as the Herzberg–Teller coupling,[78] the transition becomes allowed. Therefore, vibrational bending excitation plays an important role in the OCS dissociation dynamics. The bending excitation in OCS increases the FC overlap between the ground electronic state and the excited states at the lower excitation energy region; it reduces the energy difference between the energy surfaces and enhances the potential curve crossing.[39,49–51,53,55,58,59]

5.2.2 VMP of OCS in the Second and Higher Absorption Bands

Although most VMP studies on OCS were carried out in the first absorption band (see below), the first VMP studies of OCS were conducted for high-lying OCS states, around 13 eV[39] as well as in the 142–170 nm region ($\sim$ 7.3–8.7 eV).[40]

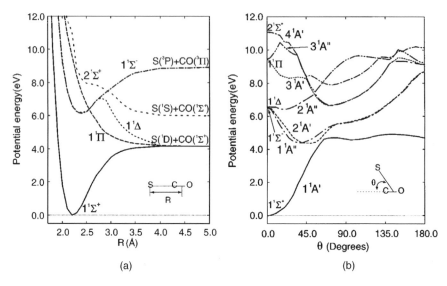

Figure 5.2 Schematics of the potential-energy curves relevant to vibrationally mediated
photodissociation of OCS (a) as a function of the internuclear distance R in
Angströms (Å) in the linear geometry with r_{co} fixed to 1.13 Å and (b) against
the bond angle θ for $R = 2.2$ Å (the equilibrium distance) and $r_{co} = 1.13$ Å.
In the first absorption band (200–250 nm, $\sim$ 5.0–6.2 eV) the $2\,^1A'(^1\Delta)$ state
carries most of the transition intensity and the $1\,^1A''(^1\Sigma^-)$ state a small
portion of it. Reprinted with permission from Ref. 81. Copyright 1998,
American Institute of Physics.

These were the first VMP studies where mode specificity was observed, although
not directly.

In the former study[39] the bending mode (v_2) of OCS was excited to either
$v_2 = 2$ by a CO_2 laser at 9.6 µm or to $v_2 = 4$ by adding a second CO_2 laser at
9.3 µm to induce the $2v_2 \rightarrow 4v_2$ transition. The vibrationally excited OCS was
dissociated by two photons of an ArF excimer laser at 193 nm:

$$OCS\left(\tilde{X}\,^1\Sigma^+, v_2\right) + 2h\nu(193\,\text{nm}) \rightarrow CO\left(X^1\Sigma^+\right) + S\left(^1S_0\right) \qquad (5.2)$$

The excited sulfur atoms were detected by observation of the collisionally
induced $XeS(2\,^1\Sigma^+ \rightarrow 1\,^1\Sigma^+)$ emission near the $S(^1S \rightarrow\,^1D)$ atomic line at
773 nm. Over 100% enhancement in the photolytic yield of $S(^1S)$ was found for
$4v_2$ pre-excitation of OCS, twice that for $2v_2$ pre-excitation. Since the photodissociation is a two-photon process, the enhancement was attributed to
mechanisms that can affect both the ground intermediates and the intermediates to final-state transitions. The suggested mechanisms are (1)
improvement of the resonance in these transitions due to the vibrational excitation, in particular for the $4v_2$ pre-excitation, (2) better FC factors for the
intermediates to final-state transitions between bent OCS states and (3) as
mentioned above, bending of ground-state OCS lower the $C_{\infty v}$ to C_s and

facilitate the ground–intermediates states transitions. The $^1\Delta$ and $^1\Pi$ states were suggested as intermediates and a high-lying $^1\Sigma$ as the final state. *However, the most interesting conclusion of this work is that a mode-specific mechanism is involved in the enhancement of the photolytic yield of* $S(^1S)$. This was deduced from the observation that the enhancement following excitation of OCS to either $2v_2$ or $4v_2$ substantially decreased when V–V vibrational equilibration was reached without greatly changing the total vibrational energy.

In the VMP of OCS in the 142–170 nm region,[40] the temperature effect on the yield of $S(^1S)$ was investigated at 232, 296 and 425 K, monitoring the $S(^1S \rightarrow {}^1D)$ emission at 773 nm. It was found that changing the temperature did not have any measurable effect on the $S(^1S)$ yield at 142–157.7 nm. In the 157.7–170 nm falloff region the yield increased with temperature, the fractional increase growing at longer wavelengths, to a factor of 1.9 at 170 nm when the temperature was changed from 232 to 425 K. Since the absorption cross section at 170 nm did not change with temperature but the population of the bending mode (v_2) at 425 K was calculated to be much higher than of all other excited vibrational modes combined, the increase was attributed, mainly, to v_2 excitation (rotational excitation might have a minor role). The main role of v_2 was in lowering the symmetry of ground-state OCS (see arguments given above for the previous example) which was apparently more effective for transitions near 170 nm.

5.2.3 VMP of OCS in the First Absorption Band

The first absorption band of OCS extends across $\sim$ 200–250 nm and peaks at 225 nm.[79] The absorption is followed by decomposition into CO molecules and S atoms:

$$OCS(\tilde{X}\,^1\Sigma^+) + h\nu \rightarrow CO(X\,^1\Sigma^+) + S(^1D_2) \text{ or } S(^3P_J) \tag{5.3}$$

where the $S(^1D_2)/S(^3P_J)$ ratio is $\sim$ 95/5 at 222 nm at room temperature and the $CO(X\,^1\Sigma^+)$ photofragments produced in the singlet channel have little vibrational excitation but are highly rotationally excited.[80] As mentioned above, most VMP works on OCS were carried out in this band[41–61] (Refs. 52,54,56,57,60 and 61, which allude only briefly to photodissociation of vibrationally excited OCS, will not be further discussed).

Some initial studies of VMP of OCS in this band were aimed at understanding the temperature dependence of photoabsorption spectra and extracting the components of the spectra related to the contributions of the populated low-energy vibrational states.[14,42,43,46] Thus, measurements of the spectral absorption coefficient of OCS around 226 nm were made at a series of temperatures from 195 to 404 K to determine whether the resulting changes in the shape of the spectrum can be attributed to contributions by specific vibrational levels of the lower electronic state.[42] Due to the relatively high frequency of the C–O stretching mode $v_1 = 1$ (2062 cm^{-1}) only the bending mode $v_2 = 1$ (520 cm^{-1}) and C–S stretching mode $v_3 = 1$ (859 cm^{-1}) were excited at the temperatures of these experiments. The results were attributed to the dependence of the electronic transition moment on vibrational excitation of the lower state, increasing significantly with excitation of the bending mode, and even more significantly with excitation of the C–S stretching mode.

This method is not specific since with increasing temperature many rovibrationally excited states are populated, complicating the direct extraction of the contribution of the different states. The complication was resolved in more elaborate studies, carried out in cold molecular beams,[49,50,53] where the specific role of the OCS bending vibrational state v_2 in the photodissociation was determined by analyzing the VMI spectra of the photofragments, CO(X $^1\Sigma$, $v = 0$, J)[49,50] and S($^1D_2$),[49,53] ionized *via* (2 + 1) REMPI. The 230-nm photodissociation dynamics starting from the v_2 (0 1 0) state was compared to that from the ground vibrational state (0 0 0): The rotational distribution of CO was measured by the (2 + 1) REMPI, in which dissociations from the two different initial states were discriminated from the translational energies of the CO fragments.[49,50] It was concluded that the transition from (0 1 0) is more than seven times stronger than the transition from (0 0 0).[50] For 288-nm dissociation the S($^1D_2$) fragment was probed. OCS population of up to $v_2 = 4$ was found in the molecular beam and it was concluded that the dissociation cross section is much higher than at 230 nm and increases by more than an order of magnitude for each step going from $v_2 = 1$ to 4 (accurate numbers could not be given since the vibrational temperature was not accurately known).[53] An additional VMP channel was deduced from these studies,[49] namely, two-photon IR (1.06 μm) excitation of the UV (230 nm) photo-prepared OCS: OCS($v = 0$) + $h\nu$(UV) → OCS*, OCS* + 2$h\nu$(IR) → OCS*(v), OCS*(v) → OCS** → CO(X $^1\Sigma^+$) + S(^1S), where the asterisks designate electronic excitation. It was suggested that the bending mode of OCS in the excited state plays a central role in the excitation and dissociation dynamics.

Other VMP studies concentrated in preparing OCS in a particular rovibrational state, using either laser excitation[41,44,47,48] or hexapole state selection,[51,55,58,59] followed by photodissociation.

In the laser-excitation studies in the first absorption band the main aim was to separate O, C or S isotopes by photodissociating selectively vibrationally excited isotopic variants of OCS (several isotopes of each of the three atoms were surveyed) and chemically scavenge the ensuing S atoms. A CO$_2$ laser at ~9.4–9.6 μm was applied to excite the 0–$2v_2$ vibrational transition of OCS.[41,44,48] Alternatively, a NH$_3$ laser at 11.4–12.3 μm, pumped by a CO$_2$ laser at 9.2 μm, was applied to pump the 0–v_1 vibrational transition of OCS.[47] In both pumping schemes the IR laser was line tuned to coincide with the absorption of specific isotopic variants and a KrF excimer laser at 249 nm was used for dissociation. Photoabsorption measurements at 249 nm were analyzed to determine the photodissociation cross section ratio $\sigma(v_2 = 0)/\sigma(v_2 = 1)/\sigma(v_1 = 1)/\sigma(v_2 = 2) = 1.0/6.2/11.0/20.5$.[47] Enrichment factors of ~ 1.5–5 were obtained for different isotopes in these experiments. The degree of isotope enrichment increased with decreasing sample temperature, due, primarily, to the decrease in the thermal population of the low vibrational levels, coupled with the strong dependence of the photodissociation cross section on vibrational excitation.[48]

Hexapole state selection in cold molecular beams of the parent molecules combined with VMI of the angular recoil of the photofragments, ionized *via* (2+1) REMPI, enabled study of state-to-state VMP of OCS ($v_2|JIM$) at ~ 230[51,55,58,59] and 223 nm.[55] It was found that the vibrational bending excitation of OCS strongly enhances the photodissociation cross section.[59] The cross section also varies with the final CO($X\ ^1\Sigma^+$, $v=0$) product rotational state. For a single J channel the maximum cross section for OCS($v_2=2$) → CO($J=64$) is a factor of 1.6 times larger than OCS($v_2=1$) → CO($J=64$) and about 33 times larger than OCS($v_2=0$) → CO($J=64$). Integrating over all J states (60–67) in the high-J region, one gets $\sigma(v_2=0)/\sigma(v_2=1)/\sigma(v_2=2)=1.0/7.0/15.0$. The enhancement of the photodissociation cross section with the bending excitation is attributed to the strong increase of the transition dipole moment with bending angle, calculated to be $0°$, $10°$ and $20°$, respectively.

5.2.4 Vector Correlations in the VMP of OCS

Several VMP studies of OCS also dealt with the anisotropy parameter β.[49,51,53,55,58] It was found that for VMP around 230 nm β depends on both the initial state of the parent and on that of the photofragment and therefore state-to-state measurements are important. At 230 nm a strong dependence of β on the initial bending state, $v_2=0$ or 1, of OCS and on the rotational state of the CO fragment was observed:[51] For OCS($v_2=0|JM=10$) the average value of β was 0.24 for CO($J=44$–55) (lower J, translationally fast) and 1.65 for CO($J=60$–65) (higher J, translationally slow); for OCS ($v_2=1|JIM=111$) the respective values of β were 0.71 and 1.74. The intermediate value of β indicates that at least two potential surfaces, namely $1\ ^1A''(^1\Sigma^-)$ and $2\ ^1A'(^1\Delta)$, are simultaneously excited and the transition dipole moment between the ground and the excited state is neither purely perpendicular, $\beta=-1$, nor parallel, $\beta=+2$, the theoretical limits for excitations to the above states, respectively. This also explains why β is largely dependent on the vibrational bending excitation of OCS, due to the difference of the angle dependence on the transition dipole moment between the two states. The transition dipole moment to the $2\ ^1A'(^1\Delta)$ state increases more steeply than that to the $1\ ^1A''(^1\Sigma^-)$ state with the bending angle of OCS.[59,81]

In a later study[58] on photodissociation of OCS $v_2=0$ or 1 around 230 nm, a dramatic decrease of β (from a large positive value to a negative value) was observed for very slow recoiling CO(J) fragments (for the same final J state, 63 or 64) when the photolysis energy was reduced by about 1000 cm^{-1}. The dramatic change of β, when strongly decreasing the recoil energy, was attributed to a breakdown of axial recoil. For VMP of OCS at 288 nm,[53] large negative values of β, -0.55 to -0.75, not dramatically changing with the degree of bending excitation, were determined as a function of the S($^1D_2$) fragment recoil speed. This indicates that the photodissociation of OCS at 288 nm occurs exclusively from the $1\,^1A''(^1\Sigma^-)$ bending excited potential surface that can be accessed through a perpendicular transition.[53]

5.3 OCSe

The dependence of $Se(^1S)$ production on the vibrational excitation of the parent molecule OCSe to $v_2=2$ by a CO_2 laser operating at 10.6 μm was examined for a photolysis wavelength of 193 nm produced by an ArF excimer laser.[62] The experimental procedure was similar to that described above for OCS^{39} except that here the two CO_2 lasers were applied for inducing different rotational transitions in the $0 \rightarrow 2v_2$ band and the vibrationally excited OCSe was dissociated by one 193-nm photon. The yield of $Se(^1S)$ was determined by monitoring the collisionally induced XeSe* emission near the $Se(^1S \rightarrow {}^1D)$ atomic line at 780 nm. Measurements of the absorption cross section for several CO_2 laser lines as well as a determination of the infrared-energy deposition for the P (34) and P (44) transitions were made. A vibrationally induced enhancement of 25% in the $Se(^1S)$ concentration was observed and interpreted as resulting from a vibrationally induced increase in the UV absorption coefficient.

5.4 N₂O

N_2O has broad photoabsorption in the 150–220 nm region, which peaks at 182 nm,[79] followed by decomposition into N_2 molecules and O atoms:[70]

$$N_2O(\tilde{X}\,^1\Sigma^+) + h\nu \rightarrow N_2\left(X\,^1\Sigma_g^+\right) + O(^1D_2) \tag{5.4}$$

Although most studies on VMP of N_2O were carried out for the above-mentioned first absorption band, a pioneering state-to-state VMP calculation was performed for the second absorption band in the 118–136 nm region, associated with the process $N_2O(\tilde{X}\,^1\Sigma) + h\nu \rightarrow N_2(X\,^1\Sigma_g^+) + O(^1S_0)$.[63] Collinear calculations were carried out, in which no artificial separation between photon absorption and the dissociation was assumed. The N_2 vibrational state distributions and absorption spectra were studied as a function of photon frequency and initial vibrational level of N_2O. The absorption spectra were found to have a single symmetric peak when starting from the (0,0,0) state and to consist of two unequal peaks when starting from the (1,0,0) (NN stretch at $2224\,cm^{-1}$) or (0,0,1) (NO stretch at $1285\,cm^{-1}$) states. If (0,0,0) is the initial state, the N_2 was mainly formed in the $v=0$ state for $\lambda > 124$ nm, while population inversion was predicted to occur in the "blue" wing ($\lambda<124$ nm) of the absorption curve. Assuming the (1,0,0) to be the initial state, the N_2 vibrational state distribution was predicted to be inverted for $\lambda<123$ nm where the absorption probability is quite significant. Starting from the (0,0,1) state, the N_2 vibrational state distributional was found to be *inverted* for most photon frequencies in which absorption occurs. Experimental examination of these results is not available.

For the first absorption band the PESs and spectroscopy of N_2O are quite similar to that of the isovalent-electronic system, OCS (both molecules are linear in the electronic ground state).[70] Also, as in OCS, in most VMP studies of N_2O the role of the bending vibrational state (0,1,0), at 589 cm^{-1},[70] in the photodissociation dynamics was investigated and similar experimental methods were applied for these studies in the two molecules (except that laser excitation of specific vibrational states was not applied in N_2O). We therefore only summarize here very briefly the main conclusions of the theoretical and experimental studies of N_2O VMP in the 150–220 nm region.[46,64–72]

Early experiments[64] and calculations[46] were aimed at understanding the temperature dependence of photoabsorption spectra and established the contribution of the v_2 bending mode to the absorption profile. Deconvolution of the experimental data showed that absorption by molecules in the (0,1,0) vibrational mode results in a spectrum of vibrational bands superimposed on a continuum.[64] Further, more elaborate theoretical[68,70,71] and experimental studies applying VMI of the photofragments,[65–67,69,72] some of them including hexapole state selection of the parent,[66,67] present more accurate figures of this contribution. In the most recent experiment reported here,[72] VMI was applied to measure the photoabsorption transition probability in the 203–205 nm region. It was estimated that the absorption probability from the first excited bending state is seven times greater than that from the ground vibrational state.

5.5 CS_2

The spectroscopy of CS_2 in the first absorption band is quite similar to that of the isovalent-electronic linear (in their ground sate) molecules OCS and N_2O although the spectrum of CS_2 is much more structured.[79] It has broad photoabsorption in the 290–350 nm region, which peaks at 320 nm[79] and leads to CS and S photoproducts. The temperature dependence of the absorption in this region was measured at five temperatures in the range 800–4000 K.[73] Since the data was taken at low resolution, a detailed analysis like that carried out on data for OCS and N_2O was not possible.[46] Nevertheless, the absorption intensity of the 320-nm system was found to increase with increasing temperature due to vibrational excitation.[73,46]

The photodissociation of CS_2 originating in various selected vibrational modes, but terminating in the same final predissociation state, was studied experimentally and the results were compared with trajectory calculations in NM coordinates.[74] Very high vibrational levels of the ground electronic state, $\tilde{X}\,^1\Sigma_g^+$, were prepared by stimulated emission pumping (SEP), using two lasers, and then excited by a third laser to the predissociative region of the $^1B_2\,^1\Sigma_u^+$ state ($\sim 5.9\,\mathrm{eV}$). By tuning the third laser, the photofragments were REMPIed and monitored by TOFMS. The complex scheme of excitation, dissociation and ionization is depicted in Figure 5.3.

16 vibrational states, from (0, 18, 0) up to (1, 26, 0) in several rotational levels were prepared as initial states. It was found that the photodissociation rate originating in combinations of the v_1 and v_2 modes is higher than that for the pure v_2 mode and there is large variation along the vibrational progression. The experimental observations agreed with the trajectory calculations.

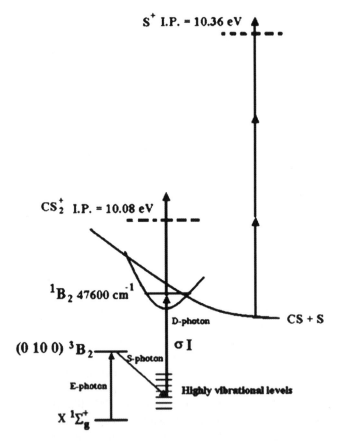

Figure 5.3   Vibrationally mediated photodissociation of CS_2 *via* stimulated emission pumping (SEP). The first laser induces the excitation of the $^3B_2$ state and the second laser populates vibrational states of the ground electronic states *via* SEP. The third laser photodissociates the molecule *via* excitation to the $^1B_2$ state and also ionizes the photofragments (as well as the CS_2 molecule). I.P.=ionization potential. Reprinted with permission from Ref. 60. Copyright 2006 American Chemical Society.

5.6 ICN

In a VMP study of ICN the change in branching between the $CN + I(^2P_{3/2})$ and the $CN + I(^2P_{1/2})$ product channels upon 248.5-nm photodissociation of vibrationally excited rather than cold ICN was measured.[74b] An increase in branching from 49% to 58% $I(^2P_{3/2})$ products was observed when the temperature of the photodissociated ICN parent was increased in going from a 100 °C to a 400 °C nozzle expansion. The crossed-laser molecular beam experiment tested a model for the dependence of branching at a conical intersection on the amplitude of the dissociative wavefunction at bent geometries. The result provided a critical comparison between the longstanding empirical surfaces for ICN photodissociation and more recent *ab initio* PESs for ICN's first absorption band. It indicated that vibrational excitation of the bending mode of the ICN parent molecules leads to increased crossing to the $I(^2P_{3/2})$ channel at the conical intersection (ICN in the ground state is linear). This observation, along with a model based on *ab initio* PESs, strongly suggests that increased branching results from increased amplitude in nonlinear C_s geometries across the conical intersection.

5.7 HCN, DCN, HOCl, DOCl and HOBr

Detailed calculations of VMP of the title molecules were carried out.[75–77] However, experimental confirmations of these calculations are not available. The photodissociation and predissociation HCN and DCN around 140 nm, yielding $H(D) + CN$ (B $^2\Sigma^+$),[75] was studied and the effect of initial vibrational excitation of the parent on the vibrational distribution of CN photoproduct was calculated. It was predicted that large isotope effects could be encountered in certain regions of the spectrum for direct photodissociation of vibrationally excited molecules.

The total photodissociation spectra and product rotational distributions resulting from $H(D)OCl(\tilde{X}$ $^1A') \rightarrow OH(D)(X$ $^2\Pi) + Cl(^2P)$, originating in a number of different vibrational states, were calculated over a continuous range of incident photon wavelengths in the range 140–440 nm.[76] It was found that the total absorption spectrum from states in which only the bending motion is excited is broader than the ground-state spectrum but similar in shape. The absorption spectrum from states in which the O–Cl stretching motion is excited show a peaked structure as a function of the wavelength of the incident light reflecting the nodal structure of the bound-state wavefunction.

The primary HOBr photodissociation products arising from absorption of radiation at 200–420 nm are OH and Br, $HOBr(\tilde{X}$ $^1A') \rightarrow OH(X$ $^2\Pi) + Br(^2P)$.[77] Predictions were made for the effect of excitation of initial vibrational states on the absorption line shapes. It was shown how the nodal pattern of the Br–OH stretching motion is reflected in the energy (or wavelength) dependence of the absorption cross section and these should be contrasted to the total absorption cross section from the (nodeless) ground vibrational state.

Excitation of the Br–OH stretching motion leads to multiple maxima in the total absorption cross section as a function of energy, while excitation of the Br–OH bending motion leads to population of higher rotational states of the OH product.

References

1. J. W. Simons, R. J. Paur, H. A. Webster III and E. J. Bair, *J. Chem. Phys.*, 1973, **59**, 1203.
2. T. E. Kleindienst and E. J. Bair, *Chem. Phys. Lett.*, 1977, **49**, 338.
3. D. H. Katayama, *J. Chem. Phys.*, 1979, **71**, 815.
4. T. E. Kleindienst, J. B. Burkholder and E. J. Bair, *Chem. Phys. Lett.*, 1980, **70**, 117.
5. P. F. Zittel and D. D. Little, *J. Chem. Phys.*, 1980, **72**, 5900.
6. I. C. McDade and W. D. McGrath, *Chem. Phys. Lett.*, 1980, **72**, 432.
7. I. C. McDade and W. D. McGrath, *Chem. Phys. Lett.*, 1980, **73**, 413.
8. S. M. Adler-Golden and J. I. Steinfeld, *Chem. Phys. Lett.*, 1980, **76**, 479.
9. D. C. Astholz, A. E. Croce and J. Troe, *J. Phys. Chem.*, 1982, **86**, 696.
10. S. M. Adler-Golden, E. L. Schweitzer and J. I. Steinfeld, *J. Chem. Phys.*, 1982, **76**, 2201.
11. J. A. Joens, J. B. Burkholder and E. J. Bair, *J. Chem. Phys.*, 1982, **76**, 5902.
12. M. G. Sheppard and R. B Walker, *J. Chem. Phys.*, 1983, **78**, 7191.
13. S. M. Adler-Golden, *J. Quant. Spectrosc. Radiat. Transfer*, 1983, **30**, 175.
14. J. A. Joens and E. J. Bair, *J. Chem. Phys.*, 1983, **79**, 5780.
15. O. Atabek, M. J. Bourgeois and M. Jacon, *J. Chem. Phys.*, 1986, **84**, 6699.
16. J. I. Steinfeld, S. M. Adler-Golden and J. Y. Gallagher, *J. Phys. Chem. Ref. Data*, 1987, **16**, 911.
17. H. A. Michelsen, R. J. Salawitch, P. O. Wennberg and J. G. Anderson, *Geophys. Res. Lett.*, 1994, **21**, 2227.
18. S. M. Ball, G. Hancock and F. Winterbottom, *Faraday Discuss.*, 1995, **100**, 215.
19. S. M. Ball, G. Hancock, J. C. Pinot de Moira, C. M. Sadowski and F. Winterbottom, *Chem. Phys. Lett.*, 1995, **245**, 1.
20. K. Takahashi, Y. Matsumi and M. Kawasaki, *J. Phys. Chem.*, 1996, **100**, 4084.
21. K. Takahashi, M. Kishigami, Y. Matsumi, M. Kawasaki and A. J. Orr-Ewing, *J. Chem. Phys.*, 1996, **105**, 5290.
22. K. Takahashi, M. Kishigami, N. Taniguchi, Y. Matsumi and M. Kawasaki, *J. Chem. Phys.*, 1997, **106**, 6390.
23. S. M. Ball, G. Hancock, S. E. Martin and J. C. Pinot de Moira, *Chem. Phys. Lett.*, 1997, **264**, 531.
24. W. Denzer, G. Hancock, J. C. Pinot de Moira and P. Tyley, *Chem. Phys. Lett.*, 1997, **280**, 496.

25. R. K. Talukdar, C. A. Longfellow, M. K. Gilles and A. R. Ravishankara, *Geophys. Res. Lett.*, 1998, **25**, 143.

26. W. Denzer, G. Hancock, J. C. Pinot de Moira and P. Tyley, *Chem. Phys.*, 1998, **231**, 109.

27. A. R. Ravishankara, G. Hancock, M. Kawasaki and Y. Matsumi, *Science*, 1998, **280**, 60.

28. K. Takahashi, N. Taniguchi, Y. Matsumi, M. Kawasaki and M. N. R. Ashfold, *J. Chem. Phys.*, 1998, **108**, 7161.

29. P. O'Keeffe, T. Ridley, K. P. Lawley, R. J. Maier and R. J. Donovan, *J. Chem. Phys.*, 1999, **110**, 10803.

30. N. Taniguchi, K. Takahashi, Y. Matsumi, S. M. Dylewski and P. L. Houston, *J. Chem. Phys.*, 1999, **111**, 6350.

31. G. Hancock and P. Tyley, *Phys. Chem. Chem. Phys.*, 2001, **3**, 4984.

32. P. O'Keeffe, T. Ridley, K. P. Lawley and R. J. Donovan, *J. Chem. Phys.*, 2001, **115**, 9311.

33. Y. Matsumi and M. Kawasaki, *Chem. Rev.*, 2003, **103**, 4767.

34. Z.-W. Qu, H. Zhu, M. Tashiro, R. Schinke and S. C. Farantos, *J. Chem. Phys.*, 2004, **120**, 6811.

35. Z.-W. Qu, H. Zhu, S. Yu. Grebenshchikov, R. Schinke and S. C. Farantos, *J. Chem. Phys.*, 2004, **121**, 11731.

35(a). H. Zhu, Z.-W. Qu, S. Yu. Grebenshchikov, R. Schinke, J. Malicet, J. Brion and D. Daumont, *J. Chem. Phys.*, 2005, **122**, 024310.

36. E. Baloitcha and G. G. Balint-Kurti, *Phys. Chem. Chem. Phys.*, 2005, **7**, 3829.

37. S. J. Horrocks, P. J. Pearson and G. A. D. Ritchie, *J. Chem. Phys.*, 2006, **125**, 133313.

38. S. J. Horrocks, G. A. D. Ritchie and T. R. Sharples, *J. Chem. Phys.*, 2007, **126**, 044308.

38(a). S. Yu. Grebenshchikov, Z.-W. Qu, H. Zhu and R. Schinke, *Phys. Chem. Chem. Phys.*, 2007, **9**, 2044.

38(b). R. Schinke and S. Yu. Grebenshchikov, *Phys. Chem. Chem. Phys.*, 2007, **9**, 4026.

39. D. J. Kligler, H. Pummer, W. K. Bischel and C. K. Rhodes, *J. Chem. Phys.*, 1978, **69**, 4652.

40. G. Black and R. L. Sharpless, *J. Chem. Phys.*, 1979, **70**, 5567.

41. P. F. Zittel and L. A. Darnton, *J. Chem. Phys.*, 1982, **77**, 3464.

42. J. R. Locker, J. B. Burkholder, E. J. Bair and H. A. Webster III, *J. Phys. Chem.*, 1983, **87**, 1864.

43. J. A. Joens and E. J. Bair, *J. Phys. Chem.*, 1983, **87**, 4614.

44. P. F. Zittel, L. A. Darnton and D. D. Little, *J. Chem. Phys.*, 1983, **79**, 5991.

45. J. A. Joens and E. J. Bair, *J. Phys. Chem.*, 1984, **88**, 6009.

46. J. A. Joens, *J. Phys. Chem.*, 1985, **89**, 5366.

47. P. F. Zittel and D. E. Masturzo, *J. Chem. Phys.*, 1986, **85**, 4362.

48. P. F. Zittel and V. I. Lang, *J. Photochem. Photobiol. A: Chem.*, 1991, **56**, 149.

49. A. Sugita, M. Mashino, M. Kawasaki, Y. Matsumi, R. Bersohn, G. Trott-Kriegeskorte and K.-H. Gericke, *J. Chem. Phys.*, 2000, **112**, 7095.
50. H. Katayanagi and T. Suzuki, *Chem. Phys. Lett.*, 2002, **360**, 104.
51. A. J. van den Brom, T. P. Rakitzis, J. van Heyst, T. N. Kitsopoulos, S. R. Jezowski and M. H. M. Janssen, *J. Chem. Phys.*, 2002, **117**, 4255.
52. T. P. Rakitzis, A. J. van den Brom and M. H. M. Janssen, *Science*, 1852, **303**, 2004.
53. M. H. Kim, W. Li, S. K. Lee and A. G. Suits, *Can. J. Chem.*, 2004, **82**, 880.
54. A. J. van den Brom, T. P. Rakitzis and M. H. M. Janssen, *J. Chem. Phys.*, 2004, **121**, 11645.
55. A. J. van den Brom, T. P. Rakitzis and M. H. M. Janssen, *J. Chem. Phys.*, 2005, **123**, 164313.
56. A. V. Komissarov, M. P. Minitti, A. G. Suits and G. E. Hall, *J. Chem. Phys.*, 2006, **124**, 014303.
57. A. J. van den Brom, P. T. Rakitzis and M. H. M. Janssen, *Phys. Scr.*, 2006, **73**, C83.
58. M. L. Lipciuc and M. H. M. Janssen, *Phys. Chem. Chem. Phys.*, 2006, **8**, 3007.
59. M. L. Lipciuc and M. H. M. Janssen, *J. Chem. Phys.*, 2007, **126**, 194318.
60. M. Brouard, A. V. Green, F. Quadrini and C. Vallance, *J. Chem. Phys.*, 2007, **127**, 084304.
61. M. Brouard, F. Quadrini and C. Vallance, *J. Chem. Phys.*, 2007, **127**, 084305.
62. W. K. Bischel, J. Bokor, J. Dallarosa and C. K. Rhodes, *J. Chem. Phys.*, 1979, **70**, 5593.
63. M. Shapiro, *Chem. Phys. Lett.*, 1977, **46**, 442.
64. G. S. Selwyn and H. S. Johnson, *J. Chem. Phys.*, 1981, **74**, 3791.
65. D. W. Neyer, A. J. R. Heck, D. W. Chandler, J. M. Teule and M. H. M. Janssen, *J. Phys. Chem. A*, 1999, **103**, 10388.
66. M. H. M. Janssen, J. M. Teule, D. W. Neyer and D. W. Chandler, in: *Atomic and Molecular Beams, State of the Art*, ed. R. Campargue, Springer, Berlin, 2000, p. 317.
67. J. M. Teule, G. C. Groenenboom, D. W. Neyer, D. W. Chandler and M. H. M. Janssen, *Chem. Phys. Lett.*, 2000, **320**, 177.
68. M. S. Johnson, G. D. Billing, A. Gruodis and M. H. M. Janssen, *J. Phys. Chem. A*, 2001, **105**, 8672.
69. T. Nishide and T. Suzuki, *J. Phys. Chem. A*, 2004, **108**, 7863.
70. S. Nanbu and M. S. Johnson, *J. Phys. Chem. A*, 2004, **108**, 8905.
71. M. N. Daud, G. G. Balint-Kurti and A. Brown, *J. Chem. Phys.*, 2005, **122**, 054305.
72. H. Kawamata, H. Kohguchi, T. Nishide and T. Suzuki, *J. Chem. Phys.*, 2006, **125**, 133312.
73. J. E. Dove, H. Hippler, H. J. Plach and J. Troe, *J. Chem. Phys.*, 1984, **81**, 1209.

74. H. T. Liou, Y. C. Chang and Z. Liou, *J. Phys. Chem. A*, 2006, **108**, 4610.

74(a). H. Guo and G. C. Schatz, *J. Chem. Phys.*, 1990, **92**, 1634.

74(b). P. W. Kash and L. J. Butler, *J. Chem. Phys.*, 1992, **96**, 8923.

75. J. A Beswick, M. Shapiro and R. Sharon, *J. Chem. Phys.*, 1977, **67**, 4045.

75(a). M. D. Morse, Y. B. Band and K. F. Freed, *J. Chem. Phys.*, 1983, **78**, 6066.

76. A. R. Offer and G. G. Balint-Kurti, *J. Chem. Phys.*, 1994, **101**, 10416.

77. G. G. Balint-Kurti, L. Fusti-Molnar and A. Brown, *Phys. Chem. Chem. Phys.*, 2001, **3**, 702.

77(a). H. Okabe, *Photochemistry of Small Molecules*, Wiley, New York, 1978.

77(b). T. Shimanouchi, *Tables of Molecular Vibrational Frequencies Consolidated Volume I*, National Bureau of Standards, Washington D.C., 1972.

78. G. Herzberg and E. Teller, *Z. Phys. Chem. B*, 1933, **21**, 410.

79. J. W. Rabalais, J. M. McDonald, V. Scherr and S. P. McGlynn, *Chem. Rev.*, 1971, **71**, 73.

80. G. Nan, I. Burak and P. L. Houston, *Chem. Phys. Lett.*, 1993, **209**, 383.

81. T. Suzuki, H. Katayanagi, S. Nanbu and M. Aoyagi, *J. Chem. Phys.*, 1998, **109**, 5778.

CHAPTER 6

VMP of Water Isotopologues

There are more VMP studies of water isotopologues than of any other species and about sixty papers that report on or allude to these studies are listed below.[1-56a] Also, water is the first molecule where three-dimensional state-to-state VMP calculations were carried out.[1] Although VMP of both the first and the second absorption band of water (the $\tilde{A}\,^1B_1 \leftarrow \tilde{X}\,^1A_1$ transition at 140–185 nm and the $\tilde{B}\,^1A_1 \leftarrow \tilde{X}\,^1A_1$ at 125–137 nm, respectively) was investigated, most studies were on the former. Following a short presentation in Section 6.2 of the theoretical work on VMP of the second absorption band, the rest of this chapter concentrates on VMP of the first absorption band.

6.1 Spectral and Dynamical Features of Water Relevant to VMP

The photodissociation of water *via* excitation in the first and second absorption bands is schematically depicted in Figure 6.1.

Both experimental and theoretical state-to-state VMP studies for the first absorption band were carried out, the first studies are presented in Refs. 2–6. For this band the VMP scheme for H_2O is presented by:

$$H_2O(\tilde{X}\,^1A_1, \nu) + h\nu \rightarrow H_2O(\tilde{A}\,^1B_1) \rightarrow OH(X^2\Pi) + H(^2S) \qquad (6.1)$$

Similar VMP schemes apply for HOD and D_2O. $H_2O(\tilde{X}\,^1A_1)$ is bent with 104° bond angle;[33] its fundamental vibrations are $\nu_1 = 3657\,cm^{-1}$ and $\nu_3 = 3756\,cm^{-1}$ for the symmetric and antisymmetric stretch, respectively, and $\nu_2 = 1595\,cm^{-1}$ for the bending; the respective values for D_2O are $\nu_1 = 2671\,cm^{-1}$, $\nu_3 = 2788\,cm^{-1}$ and $\nu_2 = 1178\,cm^{-1}$. For HOD $\nu_1 = 2727\,cm^{-1}$ for the OD stretch, $\nu_3 = 3707\,cm^{-1}$ for the OH and $\nu_2 = 1402\,cm^{-1}$ for the bending.[56b]

Vibrationally Mediated Photodissociation
By Salman (Zamik) Rosenwaks
© Salman (Zamik) Rosenwaks 2009
Published by the Royal Society of Chemistry, www.rsc.org

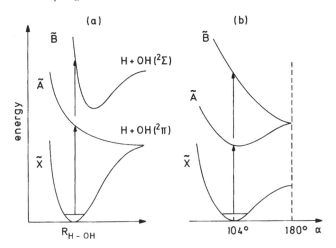

Figure 6.1   The photodissociation of water *via* excitation to the $\tilde{A}$ and $\tilde{B}$ states illustrated on one-dimensional cuts along the minimum-energy paths of the three lowest PES of H_2O. (a) The dependence of the PES on the H–OH distance. (b) The dependence of the PES on the HOH bending angle α. Reproduced from Ref. 33 with permission of Cambridge University Press.

The study of the photodissociation of the water isotopologues in the first absorption band has been attractive since it can serve as a model system for simple VMP processes.[5,33] Only one bound and one repulsive excited state are involved and for O–H internuclear distances relevant to VMP, the latter state is well isolated from all other excited electronic states.[25] Since the $\tilde{A}\,^1B_1$ excited state of H_2O is strongly repulsive, the fragmentation is fast and direct, leading exclusively to ground-state OH and H fragments. The system is small enough to enable construction of the semiempirical ground-state PES[57,57a] and to be treated by exact quantum-mechanical methods for the determination of *ab initio* potential surfaces, *e.g.*, the $\tilde{A}\,^1B_1$ surface,[58] and for treatment of the dynamics on the surfaces.[4,33,59,60] Moreover, the spectroscopy of the water molecule has been extensively studied and it is relatively easy to selectively prepare it in specific rovibrational states. These states survive the typical timescale of VMP experiments (ns). Furthermore, state-resolved detection of the ground-state photofragments, OH and H, is relatively simple. In addition, in HOD levels involving O–H and O–D stretches lie energetically rather far apart, and therefore the effect of couplings between them is negligibly small, at least for the low-lying vibrational states. This enables the study of the effect of O–H *vs.* O–D preparation on the identity of the photofragments in VMP of HOD.

In most VMP experiments on the water isotopologues the parent vibration was prepared by one-photon excitation using IR, NIR or VIS lasers, but SRE was also applied for this purpose.[16,28,29] UV lasers were used for dissociation of the vibrationally excited parent and in most experiments the OH and OD (and in one case the H[56a]) photofragments were detected by LIF, but REMPI detection of the H and D atoms was applied as well.[52] (If not otherwise stated, in the experiments described below IR/NIR/VIS vibrational excitation and OH/OD LIF detection was used).

Before going into details, a word about notation of vibrational modes is required. For the fundamental vibrations of H_2O, NM notation (v_1, v_2, v_3) = (symmetric stretch, bend, antisymmetric stretch) is mostly used. However, for the high overtones of H_2O LMs are a much better zero-order basis (expansion in either basis describes the same molecular eigenstates of the full Hamiltonian).[61–64] A LM will hereafter be denoted by $(mn^\pm, k)$, where m and n refer to the number of quanta in the OH stretches, $(+)$ and $(-)$ to symmetric and antisymmetric combinations of the states (these combinations are the degenerate eigenfunctions of the zero-order Hamiltonian), $mn^\pm = 2^{-1/2}(mn \pm nm)$, and k refer to the number of quanta in the bend. Obviously, these notations are applicable also for D_2O and HOD but for the latter the above symmetry considerations are not relevant and the degeneracy due to the dynamical tunneling effect that produces the $mn^\pm$ doublets is destroyed. Note that some authors use the notation $|mn^\pm k\rangle$ rather than $(mn^\pm, k)$ or just $|mn\rangle^\pm$ when $k = 0$.

In the LM notation the O–H stretch structure for H_2O can best be presented by polyads, with each polyad containing $v_{OH} + 1$ levels corresponding to v_{OH} stretching quanta distributed between the two identical bonds.[50,61–64] Due to anharmonic detuning effects that increase with v_{OH}, these quantum states can be described by LM excitation, e.g., $mn^\pm = 2^{-1/2}(mn \pm nm)$, with perturbative contributions from other nearby members of the same polyad, e.g., $2^{-1/2}[(m \pm n, n \mp 1) \pm (m \mp 1, n \pm 1)]$. At least for higher polyad numbers (e.g., $v_{OH} = 4$), this has led to the "spectator" paradigm, i.e. a strong propensity for cleavage of the O–H bond with greater LM vibrational excitation, with the surviving O–H bond retaining its initial LM excitation. This is exemplified below (Section 6.3.2) for the 218.5- and 239.5-nm photodissociation of polyads of $v_{OH} = 4$ in H_2O. In related examples, described in Section 6.4.2, nearly 100% selective bond fission was demonstrated in analogous HOD studies, for which the OH vs. OD stretch LM behavior was essentially complete.

6.2   VMP of H_2O and HOD in the Second Absorption Band

In the work on the second absorption band, three-dimensional quantum-mechanical calculations of the VUV photodissociation of H_2O and HOD in the ground (0,0,0) and bending (0,1,0) vibrational states were performed.[1] The dynamical equations in the ground and excited states were solved by a coupled channels expansion using the artificial channel method. The photoabsorption spectrum in the 129–136 nm range was computed. A progression of "Feshbach-type" rotational resonances, whose positions coincide very nicely with the well-known diffuse bands of water was obtained. A bimodal rotational state distribution of the $OH(X\,^2\Pi)$ photofragment was shown to exist as a result of an interplay between the direct process (giving rise to an inverted "abnormal" distribution) and a compound process (resulting in a substantial contribution of a thermal-like component). The branching ratio for OH/OD production was shown to be a sensitive function of photon energy, as were the OH versus OD rotational state distributions. Initial vibrational excitation to the (0,1,0) state was shown to increase the absolute photodissociation cross section, but to have little effect regarding the absorption line shape and rotational state distributions. It was concluded that the dynamics is dominated by final-state interactions. At present, experimental verification of the results of the calculations of VMP of water in the second absorption band is not available.

6.3 VMP of H_2O in the First Absorption Band

The first VMP state-to-state studies probing the population of individual states of ensuing photofragments and enabling crucial comparison with theory were the innovative investigations of H_2O by Andresen and coworkers.[2,3,5] Particular rovibrational states in the antisymmetric stretch of water, (0,0,1), were prepared by dye-laser Raman-shifted photons at 2.7 μm. These states were then photodissociated by a 193-nm laser beam, followed by OH photofragments LIF tagging, using a ∼308-nm UV beam from a frequency-doubled dye laser. From the resulting spectra, the populations of $OH(X\,^2\Pi)$ in different rotational, spin-orbit and Λ-doublet states were extracted showing their sensitive dependence on the parent state, with excellent agreement with the results of a simple FC model that included the rotational and electronic degrees of freedom.[5,60] These findings spurred both theoretical and experimental activity on water and its isotopologues,[7–56a] opening the arena of state-resolved studies and particularly of controlling photodissociation pathways by VMP.

6.3.1 Vibrational-State Dependence of H_2O Absorption Cross Section

Two aspects of the vibrational state dependence of H_2O absorption were studied. The first was the enhancement of the absorption from a specific rovibrational state of H_2O over that from the vibrational ground state,[5,28] and the second was the relative photodissociation cross sections of different vibrational states having approximately the same energy but different nuclear motion.[47,56a] Whereas the former aspect exemplifies the advantage of VMP in monitoring high vibrational states of the ground electronic state of water the second is instrumental in understanding the subtle differences between LMs belonging to the same polyad.

The absorption from H_2O (0,0,1) at 193 nm (monitored for a particular rotational transition) was found to be enhanced by a factor of ∼500 compared to that from the vibrational ground state.[5] Similar enhancement, ∼550, was found for H_2O (1,0,0) absorption at 193 nm.[28] The enhancement is due both to the fact that the vibrational excitation redshift the photodissociation wavelength closer to the maximum of the absorption band and to the larger FC factors.[5,28]

The ratio between the photodissociation cross sections of H_2O prepared in all stretches of the $v_{OH} = 3$ polyad, $(03^-,0)$, $(03^+,0)$, $(12^-,0)$ $(12^+,0)$, was studied at 248 nm and found to be $1/3.0/0.08/0.17$, with no dependence on the intermediate rotational state within experimental uncertainty.[47] The strong decrease in total absorption cross section between the $(03^\pm,0)$ and $(12^\pm,0)$ states was explained by a two-dimensional time-independent quantum calculation. Although the absolute values of the wavefunction amplitudes are comparable for these two pairs of intermediate levels, the $(12^\pm,0)$ states exhibit a nodal plane in the coordinate of the nondissociating OH bond length, which in a LM picture corresponds to one quantum of vibrational excitation in the spectator OH bond. When integrated over the intermediate vibrational state wavefunction, this leads to significant cancellation in the FC overlap and therefore predicts much lower UV photolysis cross sections for $(12^\pm,0)$ than $(03^\pm,0)$. The agreement between the experiment and the calculations is reasonable.[47] The 248-nm photodissociation cross sections for some rotational states of H_2O $(04^-,0)$ and $(03^-,2)$ were also measured and found to be equal.[56a]

Varying the wavelength for dissociation from high vibrational states of H_2O provides the wavelength-dependent absorption cross section for VMP of a highly vibrationally excited state.[20] Figure 6.2 depicts the measured[20] and calculated[14] relative absorption cross section for the $(04^-,0)$ state of H_2O. The agreement between the measurement (solid points) and the wavepacket propagation calculation[14] is excellent. The oscillations in the absorption curve mirrors the nodal behavior of the $(04^-,0)$ wavefunction along the dissociation path from the FC region out into the fragment channel.[14]

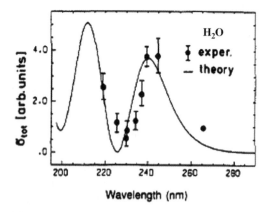

Figure 6.2   Measured (solid points)[20] and calculated (solid curve)[14] relative absorption cross section for the $(04^-,0)$ state of H_2O. The solid points are the ratio of the cross section at each wavelength to that at 266 nm. Reprinted with permission from Ref. 25. Copyright 1992 American Chemical Society.

6.3.2 Vibrational Distributions of the OH Photofragments

The influence of OH stretchings in the $H_2O(\tilde{X}\,^1A_1, v)$ parent on the vibrational distributions of the OH photofragments was experimentally studied in detail for the first through the third overtone excitation of the parent.[13,18,20,25,47,50,56a] Only OH ($v = 0$) was reported following the excitation of the fundamental vibrational stretching.[2,3,5,28] It is worth noting that theoretical studies suggest that vibrational excitation of the OH product is strong for photodissociation of H_2O (1,0,0) but weak for (0,0,1).[7] The difference was explained by a simple classical picture. For the symmetric (1,0,0) stretch there is a large initial momentum that first leads to a wide stretch of *both* OH separations and then to large-amplitude motion in the exit channel, resulting in significant vibrational excitation of the OH product. In contrast, for the antisymmetric (0,0,1) stretch the H + OH is pushed directly into the exit channel without significant OH stretch.[7]

The photodissociation of H_2O from the $(04^-,0)$ and $(13^-,0)$ stretching states, which have roughly the same energy but very different nuclear motion, was measures at 218.5, 239.5 and 266 nm.[13,18,20,25] The resulting OH vibrational distributions were interpreted as a FC mapping of the initial H_2O wavefunction, with qualitative agreement with the experiments.[14] It was found that 239.5-nm dissociation from these states leads to drastically different populations of the vibrational states of the OH photofragment: dissociation from $(04^-,0)$ produces less than 1% of OH ($v = 1$), but dissociation from $(13^-,0)$ produces a fivefold excess of OH ($v = 1$) over OH ($v = 0$).[18,20] This is believed to be due the fact that the $(04^-,0)$ wavefunction extends further along the dissociation coordinate than the $(13^-,0)$ wavefunction, but the latter has greater extension along the O–H stretching coordinate. In other words, the initial vibrational excitation in the "spectator" nondissociating O–H bond survives and primarily appears as vibrational excitation in the OH fragment for dissociation from $(13^-,0)$. The predominance of OH ($v = 0$) in the 239.5-nm dissociation from $(04^-,0)$ is consistent with the 248-nm dissociation from this state where $\sim 93\%$ of the available energy is released as relative translational energy of the nascent H + OH fragments.[56a]

For 218.5-nm dissociation, the excitation by more-energetic photons leads to regions of the excited surface nearer to the ground-state equilibrium geometry where the FC factors between bound and dissociative states is different and translation–vibration coupling and therefore vibrational excitation of the OH fragment is larger (due to *near-resonance* VMP, see further discussion of this point below and Figure 6.3). Thus, 218.5-nm photodissociation of $(04^-,0)$ and $(13^-,0)$ produces 9% and 94% of OH ($v = 1$), respectively,[18,20] and of $(03^-,1)$ and $(03^-,2)$ less than 10% of $v = 1$.[19] For 266 nm, no OH fragments were detected for $(13^-,0)$ dissociation and only OH ($v = 0$) was observed for $(04^-,0)$.[18,20] The lack of absorption and subsequent dissociation at 266 nm for $(13^-,0)$ reflects the fact that its wavefunction does not extend sufficiently far into the H + OH channel to have significant overlap with the continuum wavefunction in this case.[25]

VMP of lower polyads, $v_{OH} = 3$ and 2, at 248 and 193 nm, respectively,[47,50] showed that the OH vibrational distributions deviate considerably from the

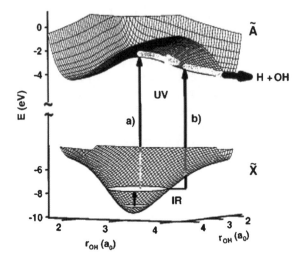

Figure 6.3   (a) On-resonance excitation where the IR and UV photons provide sufficient energy to reach the saddle point between the two H–OH and HO–H
channels on the $\tilde{A}$ state PES. The excitation is in the FC region sampling
the classically allowed portion of the initial H_2O vibrational wavefunction.
(b) Far off-resonance excitation where the energy of the IR and UV
photons is lower, the excitation is non-FC and the classically forbidden
wings are responsible for absorption. The internuclear distance r_{OH} is
given in atomic units. Reprinted with permission from Ref. 47. Copyright
1999, American Institute of Physics.

above-mentioned conventional spectator model predictions, which are based
on assuming adiabatic conservation of vibrational quanta in the surviving OH
bond.

The VMP of H_2O prepared in all stretching states of the $v_{OH} = 3$ polyad,
$(03^+,0)$, $(12^+,0)$, $(12^-,0)$ and $(03^-,0)$, was studied at 248 nm and only vibrationally *unexcited* OH products were observed despite different levels of
vibration in the spectator OH bond [(vibrational distributions are specifically
reported for $(12^-,0)$ and $(03^-,0)$].[47] Under these conditions, the total photolysis
energy is $\sim 4500\,cm^{-1}$ *below* the $\sim 55\,900\,cm^{-1}$ saddle point between the two H–
OH and HO–H channels on the $\tilde{A}$ PES, (note that the saddle point in the
multiple dimensions PES has a barrier in one particular dimension), *i.e.* sampling is on the *far off-resonance* regime. Thus, FC dissociation in this case
occurs *via* excitation significantly into the exit channel on the upper potential
surface having a significant effect on OH product vibrational distributions and
only OH $(v = 0)$ is observed despite different levels of vibration in the spectator
OH bond. This is in contrast with *near-resonance* VMP studies at higher total
energies of the $(13^-,0)$ state at 218.5 and 239.5 nm, accessing the FC region, for
which LM pre-excitation in the spectator OH bond was found to be adiabatically conserved in the photolysis event.[13,18,20,25] The present results are consistent with theoretical predictions based on a simple two-dimensional quantum

model.[47] A schematic representation of *on-resonance* and *far off-resonance* VMP is depicted in Figure 6.3.

Deviations from the spectator model were also observed in 193-nm VMP of the $v_{OH} = 2$ polyad [the $(02^-,0)$, $(02^+,0)$, $(11^+,0)$, $(01^-,2)$ states] that showed significant differences in photodissociation dynamics from symmetric $(+)$ versus antisymmetric $(-)$ vibrational states, specifically with respect to (OH $(v=1)$/(OH $(v=0)$ branching ratios.[50] Thus, photodissociation of $(11^+,0)$ and $(02^+,0)$ produced $\sim 53\%$ and $\sim 37\%$ of OH $v=1$, respectively, whereas $(02^-,0)$ produced $\sim 2\%$ of $v=1$. This observation is consistent with time-dependent wavepacket simulations[14,65] that predict significant differences in the photo-dissociation dynamics of antisymmetric and symmetric states of H_2O: Wave-packets prepared from the former states are predicted to evolve initially along the antisymmetric stretch coordinate, whereas wavepackets of the latter state have an appreciable initial component along the symmetric stretch. This dif-ference shows up as a 19-fs recurrence in the wavepacket autocorrelation function, which is found for the ground vibrational state and likewise for all symmetric states but does not occur for antisymmetric ones.[14] Physically, dis-placement on the upper surface along the symmetric stretching coordinate corresponds to motion *perpendicular* to the minimum-energy path. Such motion would predict enhanced vibrational excitation of the surviving OH bond, in good agreement with the experimental observations.[50]

6.3.3 Rotational, Spin-Orbit and Λ-Doublet State Distributions of the OH Photofragments

We start this section with a recapitulation of notation. A rotational state of H_2O is designated by J_{KaKc}, where Ka and Kc are the projections of the total angular momentum J on the a- and c-axis in the prolate and oblate symmetric top limit of H_2O. The total rotational quantum number of the OH fragment *excluding the electronic spin* is designated by N. Since $J = N + 1/2$ for $^2\Pi_{3/2}(N)$ and $J = N - 1/2$ for $^2\Pi_{1/2}(N)$[66], the respective $2J+1$ degeneracies of the two spin-orbit states are $2N+2$ and $2N$. The Λ-doublet splitting of the rotational levels of OH is due to the coupling between the rotation of the molecule and the orbital motion of the electrons. The Λ-doublet states are classified according to their symmetry in the high-N limit: A' designates an electronic wavefunction that is symmetric with reflection of the spatial coordinates of the electrons in the plane of rotation of the OH Fragment and A'' the antisymmetric case.[67]

Generally, J_{KaKc} state-selected VMP of H_2O for all initial vibrational states that were studied revealed strong oscillations in the OH-fragment quantum-state populations, in particular for low N, as a function of N for a single spin-orbit/Λ-doublet manifold. Figure 6.4 shows nascent populations for the $^2\Pi_{3/2}$ Λ-doublets (the $^2\Pi_{3/2} +$ and $-$ superscripts in the figure designate the A' and A'' Λ-doublets symmetries, respectively) obtained for a progression of increasingly OH-stretch excited polyad states, $(01^-,0)$ $(02^-,0)$,$(03^-,0)$, $(12^-,0)$ $(04^-,0)$ and $(05^-,0)$ out of $J_{KaKc} = 0_{00}$.[50] The OH rotational-state populations are relatively

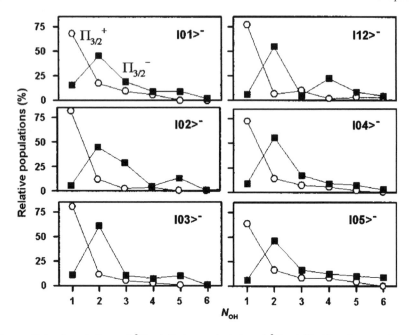

Figure 6.4 Populations of $^2\Pi_{3/2}(A')$ (open circles) and $^2\Pi_{3/2}(A'')$ (closed squares) states of OH of OH in photodissociation of different vibrational states ($v_{OH} = 1$ to 5) of H_2O in $J_{KaKc} = 0_{00}$. (The $^2\Pi_{3/2}+$ *and* $-$ superscripts in the figure designate the A' and A'' Λ-doublets symmetries, respectively). Although the photodissociation is probed at very different excess energies (the photolysis wavelength was not the same in all cases), the gross features of the rotational distributions are similar, due to the good separability of vibrational and rotational motion. Reprinted with permission from Ref. 50. Copyright 2003, American Institute of Physics.

insensitive to intermediate stretch vibrations of the same LM character, as is evident in the figure. Although the excess energies for the H_2O states in the figure vary by more than a factor of 2 (the photolysis wavelength was not the same in all cases), the N dependence of the distributions within a given spin-orbit and Λ-doublet manifold remains nearly identical.

The experiments presented below and theoretical studies demonstrated that these oscillations result from coupling between OH angular momentum states in the exit channel. In particular, a rotational FC model projecting the initial HOH wavefunction into asymptotic OH states was instrumental in describing the nascent OH distributions.[60] However, as shown below, FC models are not always sufficient and it seems that additional exit-channel interactions do play a significant role in photodissociation dynamics of H_2O at the fully state-to-state level. We note that in most cases the distributions were studied for the OH ($v = 0$) products. In the few cases where the distributions for OH ($v = 1$) were monitored, they were similar to that for OH ($v = 0$).[20,50] This indicates that the rotational, spin-orbit and Λ-doublet state distributions are uncoupled from the OH (v) vibrational state.

6.3.3.1 Rotational Distributions

The main conclusions from the results of the experiments and calculations presented below are that by and large, the initial vibrational stretching of the parent or the photodissociation wavelength has little effect on the rotational distributions, whereas the initial rotation or the bending excitation strongly influences the product rotation.

As for (0,0,1), the rotational distribution of the OH resulting from 193-nm photodissociation of the symmetric stretch of H_2O, (1,0,0), prepared by SRE, is oscillatory, depending on the particular rotation out of which the photo-dissociation is promoted,[28] and its comparison to the FC model[5,60] shows reasonable agreement. It is worth noting that the predictions of the FC theory for VMP of water were initially applied for photolysis of selected rotational states of H_2O (0,0,1).[5] Nevertheless, these calculations consider the OH rota-tional distribution as independent of the H_2O stretching motion due to the separability of the radial and angular wavefunctions of the ground and elec-tronically excited states and hence could be applied to other stretches as well. Indeed, good agreement of the OH fragment rotational distribution with the FC model was obtained also for the 218.5, 239.5 and 266 nm photolysis of water from initial rotational states of the $v_{OH} = 4$ manifold, $(04^-,0)$ and $(13^-,0)$,[13,20] the 248-nm photolysis of the $v_{OH} = 3$ manifold (but see more details on this case below), $(03^+,0)$, $(12^+,0)$, $(12^-,0)$ and $(03^-,0)$[47] and for the 193-nm photolysis of the $v_{OH} = 2$ manifold, $(02^-,0)$, $(02^+,0)$, $(11^+,0)$ and $(01^-,2)$[50] (the effect of bending in the last case is discussed below). The results of the experiments quoted above[2,3,5,13,20,28,47,50] indicate that the OH fragment rotational distribution is sensitive to the initial rotational state of the parent but *not to the initial vibrational stretching* state or photodissociation wavelength applied in these experiments.

However, the rotational state distributions resulting from the ~282-nm dissociation of the OH-stretching overtone $(04^-,0)$, $(05^-,0)$ and stretch–bend combination $(04^-,2)$,[36] differ from the above results for 193-nm dissociation from the fundamental stretches and for <282-nm photodissociation from O–H $(04^-,0)$ stretch. Qualitative agreement with model FC calculation using the theory of Ref. 60 was obtained assuming that the dissociation occurs pre-ferentially from the extended R_{H-OH} configuration,[36] R being the separation between the departing H atom and the OH center of mass (a Jacobi coordi-nate). This shows that OH rotational distributions are sensitive indicators of the *bending* and rotational motions of $H_2O(\tilde{X}\,^1A_1, v)$.

The 248-nm photolysis out of the $(03^+,0)$, $(12^+,0)$, $(12^-,0)$ and $(03^-,0)$ stretches of H_2O in the *rotational ground* state (*i.e.* $J_{KaKc} = 0_{00}$) yields OH product state distributions that demonstrate remarkable *insensitivity* to the initial choice of H_2O vibrational stretch state, in good agreement with rota-tional FC models. However, this simple trend is not followed uniformly for *rotationally excited* H_2O precursors, which indicates that these FC models are insufficient and suggests that exit-channel interactions do play a significant role in photodissociation dynamics of H_2O at the fully state-to-state level.[47]

Photodissociation from *bending* vibrations in the ground electronic state samples portions of the excited-state surface that are inaccessible from stretching vibrations and might probe the anisotropy of the excited-state surface. The influence of initial bending vibration on the rotational states of the OH product following the photodissociation of water was theoretically and experimentally studied.[4,19,25,50] It was found that excitation from an initial state with no bending excitation reaches only a rather isotropic region of the excited $H_2O(\tilde{A}\,^1B_1)$ state PES, but excitation from states with excited bending vibrations reaches more anisotropic regions.

Calculations that include the anisotropy of the excited-state potential showed that the torque exerted by the potential reduces the amount of rotational excitation below that predicted in the absence of anisotropy of this state potential. The experimental measurements of the 218.5-nm photolysis of the rotationless state $J_{KaKc} = 0_{00}$ of the $(04^-,0)$ $(03^-,1)$ and $(03^-,2)$ vibrations of H_2O, *i.e.* states having either no, one, or two quanta of bending excitation, along with four or three quanta of stretching excitation, agree with the theoretical predictions. The measured product rotation increases with increasing bending excitation, and the observed rotational excitation is less than the limit for an isotropic potential.[19] The effect of bending on the OH rotational distribution is also clearly demonstrated comparing the photolysis of $(02^-,0)$ and $(01^-,2)$ states,[50] the latter resulting in a higher rotational excitation. However, here again, the distribution is oscillatory and similar for the two states, despite the difference in population. It is suggested that this behavior is consistent with a FC picture, reflecting the approximate separability of the rotational-bending wavefunction for the H_2O intermediate state. The rotational wavefunction is predominantly responsible for the fast oscillations, while the bending wavefunction dictates the overall shape of the OH distribution.[50] A general conclusion from the above results[4,19,25,50] is that *unlike stretching, bending excitation strongly influences the product rotation.*

6.3.3.2   Spin-Orbit and Λ-Doublet Distributions

For the $\nu_{OH} = 1$ manifold of H_2O [2,3,5,25,28] the population ratio between the two spin-orbit states of the OH fragment, $OH[^2\Pi_{3/2}(N)]/[OH/^2\Pi_{1/2}(N)]$, was found to be roughly equal to the ratio of their statistical weights, $(N+1)/N$. However, deviations from statistical distributions were observed for the $\nu_{OH} = 4$ and 5 manifolds[36] (where a slight preference for $^2\Pi_{3/2}$ was found for high N values, 3–5) and for the $\nu_{OH} = 2$ and 3 manifolds.[47,50] At present, theoretical understanding of these deviations is missing.

The Λ-doublet splitting of the rotational levels of OH was monitored in almost all VMP studies of H_2O. For high OH rotational levels $\Pi(A'')$ preference is expected due to the conservation of the electronic symmetry in the break up of the excited transition complex: The $\tilde{A}\,^1B_1$ state of the dissociating molecule is antisymmetric; the hydrogen atom is formed in the $^2S_{1/2}$ symmetric state and therefore the OH is necessarily formed in the antisymmetric A'' state. Indeed, such preference, which increased with high N, was encountered in VMP of H_2O, however, parent rotations (*e.g.*, when VMP is conducted at room temperature) and complex dependence on N, often obscured this pre-ference.[2,3,5,25,28,34,36,47,50] In VMP of the $v_{OH} = 2$ manifold[50] the $\Pi(A'')/\Pi(A')$ ratio was found to be insensitive to the vibrational state of H_2O, in contrast to the OH rotational energy distribution that, as mentioned above (Section 6.1.3.1), strongly depends on the parent bending state. The magnitudes and phases of oscillations in this ratio were found to be consistent with a simple FC separability of bending and rotational wavefunctions, the bending wavefunc-tion influencing only the N-dependent features of the OH distribution and not preferential formation of Λ-doublet states with specific symmetry.[50]

6.3.4 Vector Correlations in the VMP of H_2O

The above-mentioned product final-state distributions obtained in state-to-state VMP experiments of H_2O provide information concerning the PES of the ground and electronically excited states and the bond-breaking mechanism. However, as mentioned in Section 2.1.4, these distributions do not entirely specify the photo-dissociation process. This is due to the anisotropy of the process in which the polarization of the electric field, **E**, defines a unique direction, relative to which all vectors describing both the parent molecule and the products can be correlated. Consequently, accomplishing better characterization of the VMP dynamics of H_2O requires measurement and analysis of the correlations among the parent transition dipole moment, **μ**, the recoil velocity vector of the photofragment, **v**, and its angular momentum, **J**. The vectorial properties in VMP of H_2O (1,0,0) were studied in detail using SRE for vibrational excitation and polarized 193-nm photons for dissociation of specific rotational states.[29,31,34,37,41] Both polarized broadband and narrowband LIF of OH was carried out. The 3_{03} state of H_2O (1,0,0), a pure in-plane rotation, and the $3_{21} + 3_{22} + 4_{14}$ state, consisting of a mixture of approximate contributions of in-plane (mainly 4_{14}) and out-of-plane (mainly $3_{21} + 3_{22}$) rotations, were prepared. The preparation of a pure out-of-plane rotation was limited by the weakness or resolvability of relevant transitions.[68] For the photodissociation of the 3_{03} state a significant rotational alignment, **μ**–**J**, was observed, $A_0^{(2)} = 0.42 \pm 0.10$ for the highest measured J, with about half of this value for the photodissociation of the mixed rotation. In addition, for photodissociation from the 3_{03} state the **μ**–**v**, **v**–**J**, and **μ**–**v**–**J** correlations were close to the expected values for an idealized orientation of the three vectors, *i.e.* **μ** of H_2O is parallel to the angular momentum of the OH photofragment and perpendicular to its velocity. These findings con-firmed the promptness and planarity of water photodissociation and demonstrated the instrumentality of highly specific preparation of rotations in determination of directional properties of the photodissociation.

6.3.5 VMP of H_2O as a Spectroscopic Tool

The application of VMP in vibrational spectroscopy of H_2O benefits from the high sensitivity of LIF in detecting the yield of the OH fragment as a function of the excitation wavelength preparing the rovibrational state of the H_2O parent.

Compared to direct absorption spectroscopy, this high sensitivity is particularly manifested in monitoring high vibrational or combination states of H_2O since the direct absorption decreases with increasing vibrational state. Moreover, when rovibrational states are prepared by overtone–overtone double-resonance[44,56] or triple-resonance laser-excitation schemes[55] it allows access to very high vibrational states and of different symmetry than is possible by direct absorption.

An example of the double-resonance version of VMP is presented in Figure 6.5. In this example, rovibrational levels in the electronic ground state of H_2O at the previously inaccessible energies above $26\,000\,cm^{-1}$ were measured.[56] This was achieved by first promoting H_2O in a particular rotational state to an intermediate

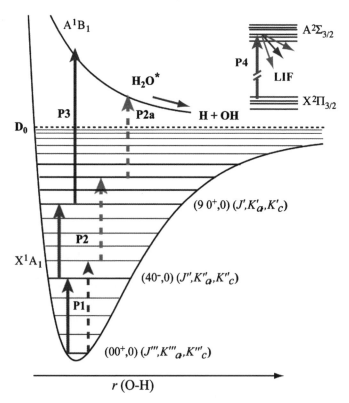

Figure 6.5   Schematic energy-level diagram of the double-resonance scheme for excitation of the OH stretch in H_2O (photons P1 and P2 shown by solid lines), with the subsequent electronic transition to the $\tilde{A}\,^1B_1$ repulsive state by a photon P3 (solid line) of 355 nm for the $(80^+,0)$, $(80^+,1)$, and $(90^+,0)$ bands or 532 nm for the $(90^+,1)$ and $(10\,0,^+0)$ bands. Excitation of the $(10\,0^+,1)$, $(11\,0^+,0)$, $(11\,0^+,1)$ and $(12\,0^+,0)$ bands (photons P1 and P2 shown by dashed lines) is followed by electronic excitation by a second photon (P2a, dashed line) of the second excitation laser. The subsequent prompt dissociation is followed by LIF detection of OH fragments. The vibrational label of the terminal state (in LM notation) is indicative only. Reprinted with permission from Ref. 56. Copyright 2007, American Institute of Physics.

level that contains four or five vibrational quanta in one of the OH stretches, then promoting a fraction of these pre-excited molecules to a higher rovibrational level containing between 8 and 12 OH-stretch quanta and between 0 and 1 quanta of OH bend and finally dissociating the prepared molecule by a UV photon and detecting the OH by LIF. The use of laser double-resonance overtone excitation extended the above limit to $34\,200\,\mathrm{cm}^{-1}$. The experimental data was used to generate a semiempirical PES that allows prediction of water levels with sub-cm^{-1} accuracy at any energy up to the above limit.[56]

In the triple-resonance laser-excitation scheme the dissociation energy of water was directly measured.[55] H_2O was excited through a series of three vibrational overtone transitions to access directly the onset of the dissociative continuum. If the terminal level prepared by the third excitation is below the dissociation threshold or if the dissociation is too slow compared to the duration of the third laser pulse, the water molecules may be further promoted to the repulsive $H_2O(\tilde{A}\,^1B_1)$ state by a second photon from the third laser, yielding OH fragments. Monitoring the OH LIF as a function of wave number of the third laser pulse in the sequence while keeping wave numbers of all other pulses fixed, generates a photofragment excitation spectrum that reflects the absorption of the overtone transition from the second intermediate state to the energy region near the dissociation threshold. Based on the *direct* observation of the onset of absorption by the OH-stretch dissociative continuum, the dissociation energy is found to be $41\,145.94 \pm 0.15\,\mathrm{cm}^{-1}$, about thirty times more accurate than the previously accepted value.[55]

6.4   VMP of HOD

The above results of the first state-to-state VMP experiments on H_2O and the theoretical predictions on bond selectivity in HOD, in the second ($\tilde{B}\,^1A_1 \leftarrow \tilde{X}$ $^1A_1)^1$ and in the first absorption band ($\tilde{A}\,^1B_1 \leftarrow \tilde{X}\,^1A_1$),[8-12] motivated experiments of controlling product identity *via* VMP of HOD. As for H_2O, most studies were carried out for the first absorption band:

$$\mathrm{HOD}(\tilde{X}\,^1A_1, \nu) + h\nu \rightarrow \mathrm{HOD}(\tilde{A}\,^1B_1) \rightarrow \mathrm{OD}(X^2\Pi) + \mathrm{H}(^2S) \qquad (6.2)$$

$$\rightarrow \mathrm{OH}(X^2\Pi) + \mathrm{D}(^2S) \qquad (6.3)$$

The photodissociation of HOD in the first absorption band was considered as a fast and direct bond-breaking process on a single PES. Quantum-mechanical time-independent[8] and time-dependent dynamics calculations[9-11] mostly used a semiempirical ground-state PES[57,57a] and an *ab initio* $\tilde{A}$ state surface.[58] These studies predicted isotopic branching ratios for the photodissociation, *via* two distinguishable channels leading to eqns (6.2) and (6.3) with small energetic differences originating from the different zero-point energies of O–D and O–H. The calculations showed that these ratios strongly depend on the frequency of the dissociating photon and the vibrational state out of which the photodissociation was initiated. The absorption curve of vibrationally excited HOD is redshifted relative to that of vibrationless ground-state H_2O, HOD and D_2O.

This shift is particularly important since VMP of HOD is usually carried out in a mixture of these three species. It affords, in favorable cases, almost exclusive production of H + OD or D + OH from vibrationally excited HOD.[9-11] Moreover, the calculations anticipated selective production of H + OD even for low vibrational excitation of OH, whereas that of D + OH only when the photodissociation begins from HOD ($3\nu_{OD}$) or higher O–D vibration.[10,11] The calculations led, already in 1990, to VMP experiments on HOD pre-excited to O–H overtones, where control of the product identity *via* VMP was shown for the first time,[15] and to O–H and O–D fundamentals.[16]

6.4.1   VMP of HOD Pre-Excited to Fundamental OH or OD Vibrations

Preparation of distinct initial vibrational eigenstates representing nearly pure O–H or O–D stretches and storing of vibrational energy exclusively in one bond or another of HOD, is a stringent test of reaction-dynamics control. VMP experiments where two different vibrational states, O–H ($\sim 3700\ cm^{-1}$) and O–D ($\sim 2700\ cm^{-1}$) stretches, served as the intermediate states from which HOD was promoted to the $\tilde{A}\ ^1B_1$ state, were reported in Refs. 16,22–24,32,39,49. In these room-temperature experiments, SRE was applied for preparing rovibrationally excited HOD molecules and CARS to monitor their excitation (see the scheme of SRE and CARS in Figure 3.3, Section 3.1.3.2). The SRE was followed by 193-nm photodissociation *via* the $\tilde{A}$ state and LIF interrogation of the OH(OD) photofragments in attempt to understand and ultimately control the course of this reaction.

Tuning of the Stokes laser through the Q-branch transitions of HOD ($1\nu_{OH}$) enabled simultaneous monitoring of the CARS spectrum (Figure 6.6(a)) and one of the reactant yield spectra, *i.e.*, action spectra of OD (Figure 6.6(b)) or OH (Figure 6.6(c)). The latter spectra were monitored while keeping the wavelength of the probe laser constant on the $R_2(4)$ line of the $A\ ^2\Sigma^+(v'=0)$ ← $X\ ^2\Pi(v''=0)$ transition of the OH or OD photofragments. Comparison of the action spectra to the CARS spectrum shows that the photodissociation of HOD molecules is enhanced whenever particular rotational states of the O–H stretch are prepared. From this enhancement and from the SRE rovibrational pumping efficiency, a ~ 300-fold increase in the 193-nm photodissociation cross section of HOD ($1\nu_{OH}$) relative to that of vibrationless ground-state water isotopologues (25% D_2O, 50% HOD and 25% H_2O present in the reaction mixture) was estimated.[22] Moreover, the OD fluorescence intensity was higher than that of the OH, demonstrating that the 193-nm photodissociation of HOD ($1\nu_{OH}$) produced more OD than OH with a branching ratio of 2.5 ± 0.5.

When similar measurements were conducted on HOD ($1\nu_{OD}$), a completely different behavior was encountered. Even though the CARS signal (Figure 6.6(d)), monitored when the SRE was tuned through the rotational levels of the HOD ($1\nu_{OD}$), seemed to be of almost similar intensity to that of $1\nu_{OH}$, neither the fluorescence of OD (Figure 6.6(e)) or OH (Figure 6.6(f)) was enhanced. These

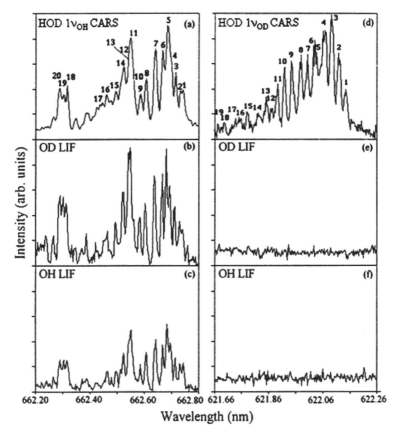

Figure 6.6 Spectra obtained in VMP of HOD ($v''_{OH} = 1$) (left panel) and ($v''_{OD} = 1$) (right panel) pre-excited by SRE and dissociated at 193 nm. The Stokes beam was scanned across the rotational levels of $v''_{OH} = 1$ and $v''_{OD} = 1$: (a) and (d) respective CARS signals; (b) and (e) LIF of the $R_2(4)$ line of $A(v' = 0) \leftarrow X(v'' = 0)$ transition of the OD photofragments, and (c) and (f) of the OH photofragments. The intensity scales in (b) and (c) are similar. The numbers above the peaks mark the Q-branch rotational transitions monitored by CARS, assigned in Table I of Ref. 22. Reprinted with permission from Ref. 22. Copyright 1991, American Institute of Physics.

measurements clearly demonstrate that excitation of even the lowest vibrational level of HOD has a dramatic effect on the photodissociation cross section. This agrees with the theoretical predictions[9–11] that the FC effects are the source of both the difference in photodissociation behavior of HOD ($1v_{OH}$) and ($1v_{OD}$) and the selectivity in bond breaking of HOD ($1v_{OH}$). The enhancement of the photodissociation cross section of HOD ($1v_{OH}$) is the result of a much better overlap of the dissociative continuum with the $1v_{OH}$ vibrational wavefunction than with either the vibrationless ground state or $1v_{OD}$. Although the vibrational excitation of the OD stretch in HOD ($1v_{OD}$) gives the O–D bond an initial "push" in the direction of the dissociation coordinate to OH + D, the unfavorable FC factor precludes the enhancement of the dissociation. In contrast, in HOD ($1\nu_{OH}$) the favorable FC factors enabled the additional energy along the OD + H dissociation coordinate to be very effective at enhancing the dissociation. Thus, these factors were the source of bond selectivity in the fragmentation of HOD ($1\nu_{OH}$), owing to the better FC overlap of the prepared wavefunction with the OD + H continuum than with the OH + D continuum.[9–11,22] This VMP study, where two distinct states with a single quantum of vibrational excitation were prepared, confirmed that it is not enough to weaken the bond to be cleaved to obtain bond selectivity, but rather a preferential FC overlap of the vibrational wavefunction with the corresponding channel on the dissociative excited state is essential.

6.4.2   VMP of HOD Pre-Excited to OH Overtones

Realization of much larger selectivity than for HOD ($1\nu_{OH}$) was obtained in the VMP experiments on HOD from states containing four[15,17,18,21,25,30,43,45] or five O–H quanta.[45] These experiments showed large enhancement of OD over OH or comparable amounts of the two, depending on the UV photon promoting the photodissociation.

VMP of $4\nu_{OH}$ photolyzed by 266-nm[21,43] photons formed at least 15-fold excess of OD, preferentially cleaving the O–H bond; similar behavior was encountered in the 239.5-nm[21,43] and 248-nm[56a] photodissociation of $4\nu_{OH}$, but this behavior changed dramatically for a still more-energetic photon of 218.5 nm producing comparable amounts of OD and OH.[21,43] These illuminating results are depicted in Figure 6.7.

Wavepacket-propagation calculations clarified the origin of the selectivity and its energy dependence in the VMP of HOD ($4\nu_{OH}$).[21,43] As indicated in Figure 6.8, the accessible regions on the upper electronic states are the keys to the dramatic change in the branching ratios for the different wavelengths. The two longer wavelengths (266 and 239.5 nm, the latter is alluded to in the figure) photons connect the $4\nu_{OH}$ vibrationally excited state to the electronically excited state in regions well out in the exit channel for breaking the O–H bond, and the total energy is below the saddle point separating the OD + H and the OH + D channels. In contrast, for the shortest wavelength either bond is broken since the energy is above the saddle point.[21,43]

A dominant OD + H channel was also apparent in the 288-nm photodissociation of rotationally state-selected HOD in the fourth and third O–H stretching overtones, with a OD + H/OH + D ratio of > 23 for the higher overtone and > 12 for the lower.[45] The reason for the decrease in the branching ratio with decreasing parent OH-stretch excitation can be understood from Figure 6.9 (in particular in the inset). Excitation from $5\nu_{OH}$ to the $\tilde{A}$ state should occur from near the maximum amplitude of the vibrational wavefunction, close to the outer classical turning point. In contrast, excitation from $4\nu_{OH}$ should occur preferentially from a classically forbidden region of the bound-state potential, where the vibrational wavefunction drops off exponentially from a peak close to the classical turning point.

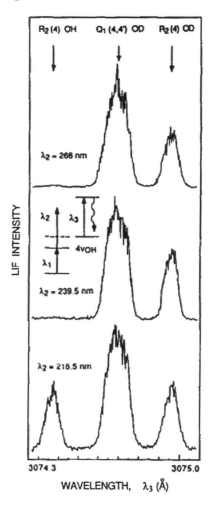

Figure 6.7 LIF excitation spectra of the OH and OD products of VMP of HOD. The vibrational excitation laser wavelength (λ_1) excites four quanta of O–H stretch ($4\nu_{OH}$) in HOD, the photolysis-laser wavelength (λ_2) dissociates the molecule, and the probe-laser wavelength (λ_3) detects either OH or OD fragments. The arrows mark the wavelengths for the LIF excitation of the OH or OD fragments. The three spectra from top to bottom are for photolysis with $\lambda_2 = 266$, 239.5, and 218.52 nm, respectively. Reprinted with permission from Ref. 21. Copyright 1991, American Institute of Physics.

Figure 6.9 presents all VMP experiments in HOD where the OH stretch was pre-excited. Note that the 288-nm photodissociation of HOD from $4\nu_{OH}$ and $5\nu_{OH}$ provides ~ 8000 and $\sim 11\,000\,\text{cm}^{-1}$ excess energy, respectively, for partitioning into the photofragments.[45] This is considerably lower than that provided in the 218.5, 239.5 and 266 nm,[21] and 248 nm[56a] experiments on HOD ($4\nu_{OH}$) and the 193-nm experiments on HOD ($1\nu_{OH}$).[22] Simple energy-conservation arguments suggest that excitation to the $\tilde{A}$ state surface in the 288-nm experiments

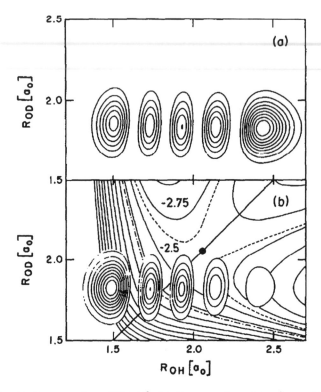

Figure 6.8 (a) Contour plot of $[(04,0)]^2$. (b) Contour plot of the $\tilde{A}$ state PES of HOD for a bending angle of 104°. The internuclear distances R_{OH} and R_{OD} are given in atomic units. The energy contours are in eV, with $E=0$ corresponding to $H+O+D$, and a spacing of 0.25 eV. The thick line indicates the ridge on the $\tilde{A}$ PES and the heavy dot marks the saddle point. The contour denoted by a dashed line (---) marks the excitation energy of the (04,0) state (the $4\nu_{OH}$ stretch) of HOD at $\lambda_2 = 238.5$ nm and by a dot-dashed line (-·-) at 218.5 nm. Superimposed are contours of $[\mu(04,0)]^2$ where μ is the $\tilde{A} \leftarrow \tilde{X}$ transition dipole moment. Note the nodal structure of the vibrational wavefunctions and the decrease of the transition moment with increasing H–OD distance. Reprinted with permission from Ref. 21. Copyright 1991, American Institute of Physics.

should occur preferentially from larger R_{H-OD} internuclear separations of the ground state than in the other experiments. The total energy for the 288-nm VMP is, like that for 266, 239.5 and 193 nm, below the saddle point separating the $OD+H$ and the $OH+D$ channels.

The rotational distributions of the OD fragments from the 288-nm VMP of HOD $4\nu_{OH}$ and $5\nu_{OH}$ was measured and accounted for by a modified FC model.[45] The FC calculations suggest that the OD photofragment rotational state distributions reflect primarily the rotational and zero-point bending motions in the selected rovibrational state of HOD. However, in contrast to the state-selected photodissociation of H_2O, the OD rotational distributions from HOD also reveal the influence of a small exit-channel torque, which is enhanced

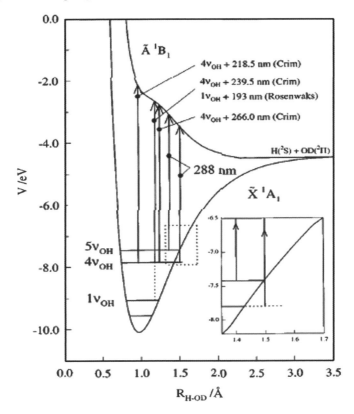

Figure 6.9 A plot of HOD ground-state potential and the $\tilde{A}$ state potential against the dissociation coordinate R_{H-OD}. Excitation to the excited-state potential at 288 nm[45] should preferentially occur from the more extended H–OD configuration than at 266, 239.5 and 218.5 nm[21] or at 193 nm.[22] The line for 248-nm $4\nu_{OH}$ dissociation,[56a] that should be squeezed between the 239.5 and 266 nm lines, is not shown in the drawing. Reprinted with permission from Ref. 45. Copyright 1997, American Institute of Physics.

in the photodissociation of the deuterated molecule by the shift in the OD center-of-mass relative to that in OH.

To conclude this section, an additional point should be raised concerning the agreement between the measured (OD + H)/(OH + D) branching ratios and the calculations. In the VMP studies of $4\nu_{OH}$ and $5\nu_{OH}$[21,45] and in the 157-nm photolysis of ground-state HOD[69] the agreement was very good. However, in the 193-nm photodissociation of $1\nu_{OH}$[22] and $3\nu_{OD}$ (see below)[40] and ground-state HOD,[46] the measured ratio were smaller than expected, with less-satisfying agreement. It was suggested that another mechanism, which primarily affects the photodissociation in the long-wavelength region of the first absorption band of water, has to be considered.[46,70] A possible mechanism may be related to contributions to the photodissociation in the long-wavelength region, *i.e.*, non-FC region, from both the singlet, $\tilde{A}$ $^1B_1$ state and a lowest triplet PES, $^3B_1$.[46,70] This indicates that even the photolysis of the benchmark molecule is not yet fully understood, in the red tail region of the absorption, and points to the importance of the interplay between theory and experiment to gain a better understanding of this subject.

6.4.3   VMP of HOD Pre-Excited to OD Overtones

As mentioned above, fragmentation to OH + D is not easily achieved, but can be accomplished after preparation of HOD ($3\nu_{OD}$) or higher O–D vibration before the addition of the UV photon.[10,11] Indeed, in direct excitation of $3\nu_{OD}$ by the Raman-shifted output of a dye laser, around $7960\,cm^{-1}$, followed by 193-nm photodissociation and LIF detection of OH or OD, bond cleavage in favor of the OH + D channel was observed, with a (OH + D)/(OD + H) branching ratio of 2.6 ± 0.5.[40] Here also, the origin of the selectivity is the favorable FC overlap that enables the additional rovibrational energy to effectively enhance the dissociation and particularly to be channeled along the OH + D channel. It is worth noting that the observed ratios for photolysis of both HOD ($1\nu_{OH}$) and HOD ($3\nu_{OD}$) agree only qualitatively with the calculated ones.[9–11,22,40]

Furthermore, (OH + D)/(OD + H) branching ratio > 12 was detected from vibrationally excited HOD in the $5\nu_{OD}$ state, photolyzed at 243.1 nm.[52] Unlike most of the previous studies of VMP of water isotopologues cited above, where the molecular photofragments where detected by LIF, the branching ratio was determined here by monitoring the H or D atomic photofragments utilizing (2 + 1) REMPI at ∼243.1 nm. While the D atom could be monitored, the H-atom REMPI signal was too weak to be detected. The present experiment was the first observation of highly selective OD dissociation of HOD with almost complete selectivity. The observed high selectivity could be explained from the spatial distribution of the vibrational wavefunction and the energetics as shown in Figure 6.10 and explained in the figure caption. Time-dependent wavepacket calculations indicated that highly selective OD dissociation is achieved by limiting E_{total}, the total excitation energy, to less than E_{saddle}, the energy gap between the vibrational ground state in the $\tilde{X}$ state and the saddle point on the $\tilde{A}$ state PES, and due to the strong LM character of the $5\nu_{OD}$ state.[52]

Additional observations in the present experiment are the recoil velocity and angular distribution of the ejected D atom that were roughly evaluated by analyzing the line shape of the D-atom REMPI spectrum. The analysis indicated two features of the ejected D atom: strong anisotropy and large velocity. From these features it is concluded that the OD dissociation occurs with the lifetime of photoexcited HOD in the $\tilde{A}$ state being much shorter than the rotational period of ∼300 fs and that 0.95 of the excess energy is used as translational energy between the D and OH fragments.

At the end of our presentation of the VMP of HOD, it is worth mentioning some of the theoretical studies on VMP of HOD utilizing IR-UV fs shaped

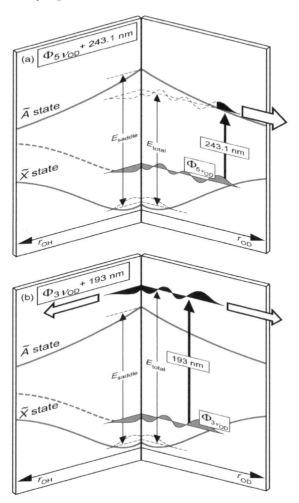

Figure 6.10   (a) Schematic diagram for 243.1-nm photolysis of HOD in the $5\nu_{OD}$ state.[52] Completely selective OD dissociation is expected since the $\tilde{A}$ state is energetically accessible only at the large OD distance (shaded area). E_{total} is the total excitation energy given by NIR + UV absorption and E_{saddle} is the energy gap between vibrational ground state in the $\tilde{X}$ state and the saddle point on the $\tilde{A}$ state PES. (b) Schematic diagram for 193-nm photolysis of HOD in the $3\nu_{OD}$ state.[40] The FC region covers over the saddle point on the $\tilde{A}$ state PES (shaded area) because of E_{total} being larger than E_{saddle}. Thus, the OH dissociation possibly proceeds in addition to the OD dissociation. Reprinted with permission from Ref. 52. Copyright 2005, American Institute of Physics.

laser pulses (although they are beyond the scope of this monograph) where control of the OH/OD branching ratio was shown.[26,49b,53–54] It was suggested that preferential O–D bond breaking in VMP of HOD ($1\nu_{OD}$) and ($2\nu_{OD}$) could be achieved by applying shaped fs-laser pulses.[53,54]

6.5 VMP of D_2O

Measurements of VMP of D_2O were reported only in one publication.[38] Rovibrationally excited D_2O in the $(03^-,0)$ state was prepared by IR excitation at $\sim 7900\,cm^{-1}$ using the Raman-shifted output of a tunable dyes laser. It was then photolyzed *via* the $\tilde{A}\,^1B_1$ state by 193-nm photons and the OD fragment detected by LIF. The $\approx 1\,eV$ vibrational excitation of D_2O led to a five orders of magnitude increase in the absorption cross section as compared to 193-nm photolysis of vibrationless D_2O, and the subsequent photolysis produced OD with $(v=1)/(v=0) = 0.39 \pm 0.07$. Based on energy considerations, the significant population of OD $(v=1)$, 28%, in the VMP of D_2O can be compared to the 218.5-nm VMP of the $(04^-,0)$ state of H_2O discussed above (Section 6.3.2), where 9% of OH $(v=1)$ is produced.[18,20] The excess energy in the photolysis (the combined energies of the vibrational overtone and photolysis photon minus the bond dissociation energy) of D_2O, $\sim 18\,470\,cm^{-1}$, is close to that in the comparable photolysis of H_2O, $\sim 18\,380\,cm^{-1}$. In both cases this behavior is consistent with the picture presented above (6.3.2) that the more-energetic photons excite the molecule from the initially prepared vibrational state to regions of the excited PES nearer to the ground-state equilibrium geometry where the translation–vibration coupling is larger. Apparently, there is a better coupling in the D_2O case.

References

1. E. Segev and M. Shapiro, *J. Chem. Phys.*, 1982, **77**, 5604.
2. P. Andresen, V. Beushausen, D. Häusler and H. W. Lülf, *J. Chem. Phys.*, 1985, **83**, 1429.
3. R. Schinke, V. Engel, P. Andresen, D. Häusler and G. G. Balint-Kurti, *Phys. Rev. Lett.*, 1985, **55**, 1180.
4. R. Schinke, V. Engel and V. Staemmler, *J. Chem. Phys.*, 1985, **83**, 4522.
5. D. Häusler, P. Andresen and R. Schinke, *J. Chem. Phys.*, 1987, **87**, 3949.
6. P. Andresen and R. Schinke, in *Advances in Gas-Phase Photochemistry and Kinetics: Molecular Photodissociation Dynamics*, ed. M. N. R. Ashfold and J. E. Bagott, Royal Society of Chemistry, London, 1987, p. 61.
7. V. Engel, R. Schinke and V. Staemmler, *J. Chem. Phys.*, 1988, **88**, 129.
8. V. Engel and R. Schinke, *J. Chem. Phys.*, 1988, **88**, 6831.
9. J. Zhang and D. G. Imre, *Chem. Phys. Lett.*, 1988, **149**, 233.
10. J. Zhang, D. G. Imre and J. H. Frederick, *J. Phys. Chem.*, 1989, **93**, 1840.
11. D. G. Imre and J. Zhang, *Chem. Phys.*, 1989, **139**, 89.
12. B. Hartke, J. Manz and J. Mathis, *Chem. Phys.*, 1989, **139**, 123.
13. R. L. Vander Wal and F. F. Crim, *J. Phys. Chem.*, 1989, **93**, 5331.
14. K. Weide, S. Hennig and R. Schinke, *J. Chem. Phys.*, 1989, **91**, 7630.
15. R. L. Vander Wal, J. L. Scott and F. F. Crim, *J. Chem. Phys.*, 1990, **92**, 803.

16. I. Bar, Y. Cohen, D. David, S. Rosenwaks and J. J. Valentini, *J. Chem. Phys.*, 1990, **93**, 2146.

17. F. F. Crim, *Science*, 1990, **249**, 1387.

18. F. F. Crim, M. C. Hsiao, J. L. Scott, A. Sinha and R. L. Vander Wal, *Philos. Trans. Roy. Soc.*, 1990, **332**, 259.

19. R. Schinke, R. L. Vander Wal, J. L. Scott and F. F. Crim, *J. Chem. Phys.*, 1991, **94**, 283.

20. R. L. Vander Wal, J. L. Scott and F. F. Crim, *J. Chem. Phys.*, 1991, **94**, 1859.

21. R. L. Vander Wal, J. L. Scott, F. F. Crim, K. Weide and R. Schinke, *J. Chem. Phys.*, 1991, **94**, 3548.

22. I. Bar, Y. Cohen, D. David, T. Arusi-Parpar, S. Rosenwaks and J. J. Valentini, *J. Chem. Phys.*, 1991, **95**, 3341.

23. Y. Cohen, D. David, T. Arusi-Parpar, I. Bar, S. Rosenwaks and J. J. Valentini, in *Mode Selective Chemistry*, eds. J. Jortner, R. D. Levine and B. Pullman, Kluwer, Dordrecht, 1991, p. 227.

24. I. Bar, Y. Cohen, D. David, T. Arusi-Parpar, S. Rosenwaks and J. J. Valentini, *J. Physique IV*, 1991, **C7**, 651.

25. V. Engel, V. Staemmler, R. L. Vander Wal, F. F. Crim, R. J. Sension, B. Hudson, P. Andresen, S. Hennig, K. Weide and R. Schinke, *J. Phys. Chem.*, 1992, **96**, 3201.

26. B. Amstrup and N. Henriksen, *J. Chem. Phys.*, 1992, **97**, 8285.

27. M. Shapiro, in *Isotope Effects in Gas-Phase Chemistry*, ed. J. A. Kaye, ACS Symposium Series 502, ACS, Washington D.C., 1992, p. 264.

28. D. David, A. Strugano, I. Bar and S. Rosenwaks, *J. Chem. Phys.*, 1993, **98**, 409.

29. D. David, I. Bar and S. Rosenwaks, *J. Chem. Phys.*, 1993, **99**, 4218.

30. F. F. Crim, *Annu. Rev. Phys. Chem.*, 1993, **44**, 397.

31. D. David, I. Bar and S. Rosenwaks, *J. Phys. Chem.*, 1993, **97**, 11571.

32. S. Rosenwaks, T. Arusi-Parpar, I. Bar, P. Bouchardy, Y. Cohen, D. David, F. Grisch, D. Heflinger, M. Pealat, A. Strugano, J. P. Taran and J. J. Valentini, *Nonlin. Opt.*, 1993, **5**, 33.

33. R. Schinke, *Photodissociation Dynamics*, Cambridge University Press, Cambridge, 1993.

34. D. David, I. Bar and S. Rosenwaks, *J. Photochem. Photobiol. A: Chem.*, 1994, **80**, 23.

35. D. F. Plusquellic, O. Votava and D. J. Nesbitt, *J. Chem. Phys.*, 1994, **101**, 6356.

36. M. Brouard, S. R. Langford and D. E. Manolopoulos, *J. Chem. Phys.*, 1994, **101**, 7458.

37. I. Bar, D. David and S. Rosenwaks, *Chem. Phys.*, 1994, **187**, 21.

38. Y. Cohen, I. Bar and S. Rosenwaks, *Chem. Phys. Lett.*, 1994, **228**, 426.

39. T. Arusi-Parpar, Y. Cohen, D. David, I. Bar and S. Rosenwaks, *J. Physique IV*, 1994, **C4**, 725.

40. Y. Cohen, I. Bar and S. Rosenwaks, *J. Chem. Phys.*, 1995, **102**, 3612.

41. D. David, I. Bar and S. Rosenwaks, *Nonlin. Opt.*, 1995, **11**, 13.

42. K. M. Christoffel and J. M. Bowman, *J. Chem. Phys.*, 1996, **104**, 8348.

43. F. F. Crim, *J. Phys. Chem.*, 1996, **100**, 12725.

44. R. J. Barnes, A. F. Gross and A. Sinha, *J. Chem. Phys.*, 1997, **106**, 1284.

45. M. Brouard and S. R Langford, *J. Chem. Phys.*, 1997, **106**, 6354.

46. D. F. Plusquellic, O. Votava and D. J. Nesbitt, *J. Chem. Phys.*, 1998, **109**, 6631.

47. O. Votava, D. F. Plusquellic and D. J. Nesbitt, *J. Chem. Phys.*, 1999, **110**, 8564.

48. O. Votava, D. F. Plusquellic, T. L. Myers and D. J. Nesbitt, *J. Chem. Phys.*, 2000, **112**, 7449.

49. I. Bar and S. Rosenwaks, *Int. Rev. Phys. Chem.*, 2001, **20**, 711.

49(a). R. van Harrevelt and M. C. van Hemert, *J. Chem. Phys.*, 2001, **114**, 9543.

49(b). N. Elghobashi, P. Krause, J. Manz and M. Opel, *Phys. Chem. Chem. Phys.*, 2003, **5**, 4806.

50. S. A. Nizkorodov, M. Ziemkiewicz, T. L. Myers and D. J. Nesbitt, *J. Chem. Phys.*, 2003, **119**, 10158.

51. S. A. Nizkorodov, M. Ziemkiewicz, D. J. Nesbitt and A. E. Knight, *J. Chem. Phys.*, 2005, **122**, 194316.

52. H. Akagi, H. Fukazawa, K. Yokoyama and A. Yokoyama, *J. Chem. Phys.*, 2005, **123**, 184305.

53. M. Sarma, S. Adhikari and M. K. Mishra, *Chem. Phys. Lett.*, 2006, **420**, 321.

54. S. Adhikaria, S. Deshpandea, M. Sarma, V. Kurkala and M. K. Mishra, *Radiat. Phys. Chem.*, 2006, **75**, 2106.

55. P. Maksyutenko, T. R. Rizzo and O. V. Boyarkin, *J. Chem. Phys.*, 2006, **125**, 181101.

56. P. Maksyutenko, J. S. Muenter, N. F. Zobov, S. V. Shirin, O. L. Polyansky, T. R. Rizzo and O. V. Boyarkin, *J. Chem. Phys.*, 2007, **126**, 241101.

56(a). A. Läuter, P. D. Naik, J. P. Mittal, H-R. Volpp and J. Wolfrum, *Res. Chem. Intermed.*, 2007, **33**, 513.

56(b). T. Shimanouchi, *Tables of Molecular Vibrational Frequencies Consolidated Volume I*, National Bureau of Standards, Washington D.C., 1972.

57. K. S. Sorbie and J. N. Murrell, *Mol. Phys.*, 1975, **29**, 1387.

57(a). J. R. Reimers and J. R. Watts, *Mol. Phys.*, 1984, **52**, 357.

58. V. Staemmler and A. Palma, *Chem. Phys.*, 1985, **93**, 63.

59. G. G. Balint-Kurti and M. Shapiro, *Chem. Phys.*, 1981, **61**, 137.

60. G. G. Balint-Kurti, *J. Chem. Phys.*, 1986, **84**, 4443.

61. R. T. Lawton and M. S. Child, *Mol. Phys.*, 1979, **37**, 1799.

62. R. T. Lawton and M. S. Child, *Mol. Phys.*, 1980, **40**, 773.

63. M. S. Child and R. T. Lawton, *Chem. Phys. Lett.*, 1982, **87**, 217.

64. M. S. Child and L. Halonen, *Adv. Chem. Phys.*, 1984, **57**, 1.

65. J. Zhang and D. G. Imre, *J. Chem. Phys.*, 1989, **90**, 1666.

66. G. Herzberg, *The Spectra and Structure of Simple Free Radicals*, Cornell University Press, Ithaca and London, 1971.

67. M. H. Alexander, P. Andresen, R. Bacis, R. Bersohn, F. J. Comes, P. J. Dagdigian, R. N. Dixon, R. W. Field, G. W. Flynn, K-H. Gericke, E. R. Grant, B. J. Howard, J. R. Huber, D. S. King, J. L. Kinsey, K. Kleinermanns, K. Kuchitsu, A. C. Luntz, A. J. McCaffery, B. Pouilly, H. Reisler, S. Rosenwaks, E. W. Rothe, M. Shapiro, J. P. Simons, R. Vasudev, J. Wiesenfeld, C. Wittig and R. N. Zare, *J. Chem. Phys.*, 1988, **89**, 1749.

68. D. David, A. Strugano, I. Bar and S. Rosenwaks, *Appl. Spectrosc.*, 1992, **46**, 1149.

69. N. Shafer, S. Satyapal and R. Bersohn, *J. Chem. Phys.*, 1989, **90**, 6807.

70. T. Schröder, R. Schinke, M. Ehara and K. Yamashita, *J. Chem. Phys.*, 1998, **109**, 6641.

VMP of Tetratomic Molecules

The VMP theories and experiments of tetratomic molecules are obviously more complex than those for triatomics. The additional atom adds additional degrees of freedom that complicate both theory and experiment. Moreover, due to the additional degrees of freedom, IVR is expected to obscure mode selectivity. In contrast to the triatomic water molecule, where detailed calculations predicted selectivity, similar predictions were not available prior to VMP experiments on tetratomic molecules. Nevertheless, mode selectivity has been observed in several of these molecules. Extensive state-to-state VMP studies were carried out on three species, described in the next three sections in order of increasing complexity of the system: acetylene isotopologues, ammonia isotopologues and isocyanic acid. These sections are followed by a section on other tetratomic molecules, hydrogen peroxide isotopologues, nitrous acid and hydrazoic acid, where VMP was studied but mode-selective or bond-selective dissociation was not experimentally observed.

7.1 VMP of Acetylene Isotopologues

7.1.1 Spectral and Dynamical Features of Acetylene Relevant to VMP

VMP of the acetylene isotopologues C_2H_2 and C_2HD was studied at total (vibrational plus electronic) excitation energies in the range of $\sim 51\,000$–$57\,000\,cm^{-1}$.[1–14] At these energies acetylene can be photodissociated via excitation to the first and second singlet states:

$$C_2H_2\left(\tilde{X}\,^1\Sigma_g^+, v\right) + h\nu \rightarrow C_2H_2(\tilde{A}\,^1A_u \text{ or } \tilde{B}\,^1B_u)$$

$$\rightarrow C_2H(\tilde{A}\,^2\Pi) + H(^2S) \qquad (7.1)$$

$$\rightarrow C_2H\left(\tilde{X}\,^2\Sigma^+\right) + H(^2S) \qquad (7.2)$$

where similar VMP schemes apply for $C_2HD(\tilde{X}\,^1\Sigma^+, v)$.

Vibrationally Mediated Photodissociation
By Salman (Zamik) Rosenwaks
© Salman (Zamik) Rosenwaks 2009
Published by the Royal Society of Chemistry, www.rsc.org

The linear, centrosymmetric structure of the ground electronic state of C_2H_2, as well as that of the nonsymmetric C_2HD, is deceptively simple. Despite the extensive attention these species have received, their photophysical and photochemical properties are still not completely understood. The complicated spectra and dynamics are reviewed in Ref. 1, which is an authoritative summary of the vast amount of work that has been conducted on these subjects. Nevertheless, VMP of acetylene seems to be a good starting point to study VMP of tetratomic molecules since in addition to being linear in its ground electronic state, it contains a triple "rigid" bond between the carbon atoms that may hinder energy transfer between different vibrational modes that are separated by the C≡C moiety. Indeed, extensive VMP studies of both C_2H_2 and C_2HD were carried out.[2-14]

The linear, tetratomic ($N = 4$) structure of acetylene gives rise to $(3N - 5) = 7$ NMs of vibration. The modes of the $\tilde{X} \, ^1\Sigma_g^+$ ground electronic state of C_2H_2 are[15] the symmetric $\nu_1(\sigma_g^+) = 3374 \, \text{cm}^{-1}$ and antisymmetric $\nu_3(\sigma_u^+) = 3289 \, \text{cm}^{-1}$ CH stretches, the $\nu_2(\sigma_g^+) = 1974 \, \text{cm}^{-1}$ CC stretch, the $\nu_4(\pi_g) = 612 \, \text{cm}^{-1}$ *trans*-bend and the $\nu_5(\pi_u) = 730 \, \text{cm}^{-1}$ *cis*-bend; the respective values for $C_2HD(\tilde{X} \, ^1\Sigma^+)$ are the $\nu_1(\sigma^+) = 3336 \, \text{cm}^{-1}$ CH stretch, the $\nu_3(\sigma^+) = 2584 \, \text{cm}^{-1}$ CD stretch, the $\nu_2(\sigma^+) = 1854 \, \text{cm}^{-1}$ CC stretch, the $\nu_4(\pi) = 518 \, \text{cm}^{-1}$ CH bend and the $\nu_5(\pi) = 678 \, \text{cm}^{-1}$ CD bend. Each of the bending modes ν_4 and ν_5 is doubly degenerate with respective vibrational angular momentum quantum numbers l_4 and l_5. The 6 NMs of the *trans*-bend $\tilde{A} \, ^1A_u$ state of C_2H_2 are[11] the symmetric $\nu_1(a_g) = 2880 \, \text{cm}^{-1}$ CH stretch, the $\nu_2(a_g) = 1387 \, \text{cm}^{-1}$ CC stretch, the $\nu_3(a_g) = 1048 \, \text{cm}^{-1}$ *trans*-bend, the $\nu_4(a_u) = 765 \, \text{cm}^{-1}$ torsion, the $\nu_5(b_u) = 2857 \, \text{cm}^{-1}$ antisymmetric stretch and the $\nu_6(b_u) = 768 \, \text{cm}^{-1}$ *cis*-bend.

Many studies of acetylene focused on the excited vibrational states of the $\tilde{X}$ state due to their simplicity and since they represent features relevant to vibrational dynamics in larger systems.[1,16,17] This state possesses a vibrational density of states growing relatively slowly with energy, leading to fully resolved spectra up to chemically interesting levels of vibrational excitation, $E_{vib} \leq 20\,000 \, \text{cm}^{-1}$, which is well above the $\sim 15\,000 \, \text{cm}^{-1}$ barrier to a bond-breaking isomerization between acetylene and vinylidene, H_2CC.[16] The stretching and bending frequencies of the C–H oscillators occur approximately with a 5:1 ratio, leading to a weaker coupling to the rest of the molecule than in other hydrides (*e.g.*, H_2O) where this ratio is about 2:1. The monodeuterated acetylene isotopologue, C_2HD, is particularly interesting, representing an almost unperturbed energy pattern of vibrational levels,[18-20] since pairs of levels involving C–H and C–D stretches lie energetically rather far apart, reducing the effect of couplings.

The spectroscopy of the excited electronic states of acetylene[21-24] and its dissociation mechanism[25-39] has been extensively studied. These studies indicated that whereas its ground state is linear, the first two excited singlet states ($\tilde{A} \, ^1A_u$ and $\tilde{B} \, ^1B_u$) have planar *trans*-bent configurations. The $\tilde{A} \leftarrow \tilde{X}$ and $\tilde{B} \leftarrow \tilde{X}$ partially overlapping absorption bands exhibit long vibrational progressions in ν_3' (the excited state *trans*-bending mode) in the 190–240 and 155–200 nm wavelength regions, respectively. In addition, it was found that both states correlate adiabatically with excited C_2H and ground-state H $[(C_2H(\tilde{A}\ ^2\Pi) + H(^2S)]$.[27,32,34] However, the adiabatic surface has a barrier that inhibits $C_2H(\tilde{A}\ ^2\Pi)$ production and consequently enables predissociation to the ground-state products $C_2H(\tilde{X}\ ^2\Sigma^+) + H(^2S)$. The latter process is non-adiabatic and may evolve *via* intersystem crossing of $C_2H_2(\tilde{A}\ ^1A_u)$ to near-resonant triplet states and/or internal conversion (IC) to high-lying vibrational levels of the ground state. The general belief is that intersystem crossing represents the dominant coupling out of the $\tilde{A}$ state in the near-threshold energy range.[25–26,28,29,31–38]

Consideration of the above-mentioned characteristics suggests that disposal of vibrational excitation in specific modes of C_2H_2 or C_2HD might provide different starting points for photodissociation than does the photodissociation of vibrationless reactants. This can reveal entirely different aspects of the state-to-state dynamics and shed new light on this interesting molecular system. Furthermore, the possibility of obtaining isotopic selectivity in the photo-fragmentation of C_2HD may provide information concerning the nature of the prepared vibrational states and the exit-channel interactions governing the bond breaking on the initially excited bound state.

We note that as in water, the high overtones of C_2H_2 and C_2HD may be better described in a LM presentation.[40] However, since in all VMP studies NMs presentation is used, we will adopt the latter. The zero-order basis states for the vibrational levels in the $\tilde{X}$ state are labelled as $(V_1 V_2 V_3 V_4^{l_4} V_5^{l_5})^l$ where V_i designates the number of vibrational quanta in each of the NMs ($i = 1$–5), the l_4 and l_5 superscripts represent the vibrational angular momentum quantum numbers of the respective (degenerate) bending modes and $l_4 + l_5 = l$ the resultant angular momentum number.[1] The vibrational states for $l = 0,1,2, \ldots$ are designated by $\Sigma, \Pi, \Lambda, \ldots$

7.1.2   VMP of C_2H_2

VMP of C_2H_2 was carried out from vibrational levels in the $\tilde{X}$ state with energies around that of $1\nu_{CH}$,[11] $3\nu_{CH}$,[9,12,13] $4\nu_{CH}$,[2,10] and $5\nu_{CH}$[4–8] stretchings. Depending on the combined vibrational plus electronic energy, the molecule could be promoted to the $\tilde{A}$ or $\tilde{B}$ states, both correlating adiabatically with $C_2H(\tilde{A}) + H$ with an asymptote value of $\sim 49\ 690\ cm^{-1}$.[32,34] However, the adiabatic connection to $C_2H(\tilde{A})$ involves a barrier of $\sim 52\ 260\ cm^{-1}$,[32,34] confining the efficient production of $C_2H(\tilde{A})$ and consequently leading to the competing excited-state decay process, explicitly (nonadiabatically) to C_2H $(\tilde{X}\ ^2\Sigma^+) + H$ with a threshold of $\sim 46\ 540\ cm^{-1}$ (schematics of the correlations is depicted in Figure 7.5). The experiments and analysis of "adiabatic" and "nonadiabatic" VMP of C_2H_2 are presented in the following two subsections, respectively. The titles of the subsections do not indicate that the processes in the respective cases are exclusively adiabatic or nonadiabatic, but rather that the combined photodissociation energy is above or below the $\sim 52\ 260\ cm^{-1}$

barrier. We note that on the basis of lifetime measurements of $C_2H_2(\tilde{A})$ it was suggested that the barrier is around $\sim 51\,100\,cm^{-1}$ (the lifetime became longer at and above this energy).[11] However, the calculated $\sim 52\,260\,cm^{-1}$ barrier and the experiment on VMP of C_2H_2 $4\nu_{CH}$ described in the next section strongly favors the higher value. At any rate, the classification of the following sub-sections is valid also for the lower value.

It is noteworthy that it was shown that large nuclear polarizations in isolated C_2H_2 (and other molecules) may be created *via* the hyperfine interaction fol-lowing excitation to selected rovibrational states $|v,JM\rangle$.[40a] Photodissociation of vibrationless as well as vibrationally excited polarized C_2H_2 allows the production of polarized H atoms.

7.1.2.1 Adiabatic VMP of C_2H_2

(a) The $4\nu_{CH}$ Region. In the first VMP study of C_2H_2, expansion-cooled acetylene was excited to a single rotational state in the $\nu_1 + 3\nu_3$ vibrational level (4 CH stretches) and then photodissociated at 248.3 nm.[2] The HRTOF tech-nique was applied to the H photofragments, excited by sequential absorption of 121.6-nm and 366-nm photons, to obtain center-of-mass translational-energy distributions and from that the yields of $C_2H(\tilde{A})$ and $C_2H(\tilde{X})$. These dis-tributions were compared to those from one-photon 193.3-nm photodissocia-tion. As shown in Figure 7.1, the dominant product channel in the VMP experiment (a) was $C_2H(\tilde{A}) + H$, differing markedly from the 193.3-nm pho-todissociation (b), which provided $\sim 1220\,cm^{-1}$ less energy and yielded pri-marily $C_2H(\tilde{X}) + H$.

In the 248.3-nm photodissociation of $(\nu_1 + 3\nu_3)$ the combined energy $(\sim 52\,950\,cm^{-1})$ just exceeds the above-mentioned calculated barrier on the C_2H_2 $\tilde{A}$ state (the $\tilde{B}$ state is inaccessible at this energy), while in the 193.3-nm dissociation it is just below it. The obvious interpretation of the dominance of $C_2H(\tilde{A})$ in the VMP experiment is thus that overcoming the barrier due to the excess energy leads to the adiabatic $C_2H(\tilde{A}) + H$ channel. Another interpreta-tion is that moving of the FC region due to the vibrational excitation may result in suppressing the nonadiabatic $C_2H(\tilde{X}) + H$ channel.[2]

A different issue, the comparison of VMP for different vibrational levels in the $4\nu_{CH}$ region,[10] is presented in subsection *(d)* below.

(b) Comparison of VMP for Different Vibrational Levels in the $5\nu_{CH}$ Region. Extensive VMP studies were conducted in a pulsed molecular beam of C_2H_2 in the $5\nu_{CH}$ region (at $15\,480$–$15\,720\,cm^{-1}$) where excitations, by VIS pho-tons, of different bands were compared.[4-8] The vibrationally excited C_2H_2 was promoted by ~ 243.1-nm photons to the excited electronic *trans*-bent states $\tilde{A}/\tilde{B}$ and dissociated. These UV photons also interrogated the H frag-ments *via* $(2+1)$ REMPI, enabling generation of action spectra, H Doppler profiles and TOFMS. The vibrational excitation to $5\nu_{CH}$ was carried out

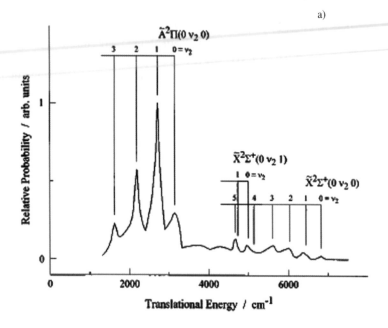

a)

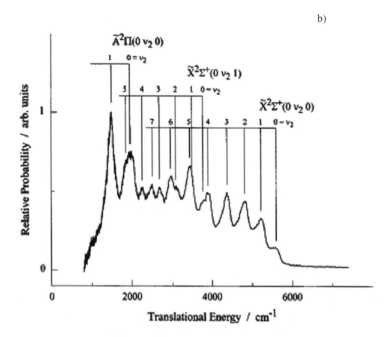

b)

Figure 7.1   Center-of-mass translational-energy distribution and assignments of C_2H product states for a) IR ($v_1 + 3v_3$) plus 248.3 nm and b) 193.3-nm photo-dissociation of C_2H_2. $C_2H(\tilde{A})$ is dominant in (a) and $C_2H(\tilde{X})$ in (b). The comb lines indicate peaks in rotational distributions of C_2H in both states. Reprinted with permission from Ref. 2. Copyright 1995, American Institute of Physics.

from both the ground and excited vibrational states. The excitation and detection scheme is depicted in Figure 7.2.

A representative action spectrum obtained in VMP of C_2H_2 (1410^02^0), (2030^00^0) and (2031^10^0) is displayed in the lower trace of Figure 7.3. The (2030^00^0) band comprises excitation of five quanta of C–H stretching, while the combination bands include C–H stretch or C–H plus C≡C stretch excitation as well as bending. The spectra of the (1410^02^0) and (2030^00^0) bands are characterized by $\Sigma_u^+ - \Sigma_g^+$ transitions including *P*- and *R*-branches with intensity

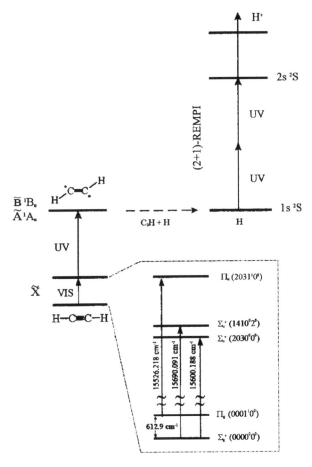

Figure 7.2   Schematics of the VMP experiments in C_2H_2 in the $5\nu_{CH}$ region. VIS wavelengths are applied for vibrational excitation and UV for the subsequent promotion to the excited electronic states $\tilde{A}\,^1A_u/\tilde{B}\,^1B_u$. The UV photons are also used to detect the H photofragments by $(2 + 1)$ REMPI. Note that the ground electronic state is linear, whereas the $\tilde{A}$ and $\tilde{B}$ states have planar *trans*-bent configurations. The inset shows the $\Sigma_u^+ - \Sigma_g^+$ and $\Pi_u - \Pi_g$ vibrational transitions and the frequencies used for the first step of excitation. Reprinted with permission from Ref. 6. Copyright 1998, American Institute of Physics.

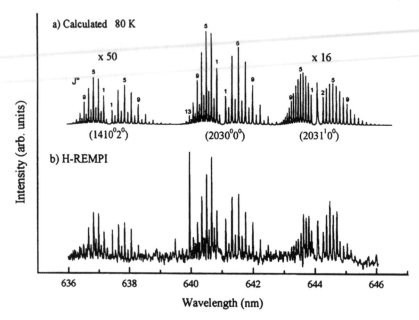

Figure 7.3 Vibrational overtone excitation spectra of the C_2H_2 $(1410^02^0) \leftarrow (0000^00^0)$, $(2030^00^0) \leftarrow (0000^00^0)$ and $(2031^10^0) \leftarrow (0001^10^0)$ transitions. The simulated absorption spectrum a) was calculated using rovibrational parameters from Refs. 41–44, nuclear spin degeneracies and the Boltzmann distribution at a rotational temperature of 80 K. The action spectrum b) presents the $(2+1)$ REMPI intensities of the H photofragments owing to 243.135-nm photolysis as a function of the visible wavelength applied for vibrational excitation. $\times 50$ and $\times 16$ are the factors used to scale up the simulated absorption intensities of the (1410^02^0) and (2031^10^0) bands relative to that of $((2030^00^0))$, respectively. The numbers above the peaks in the simulated spectrum assign the rotational levels of the vibrational ground-state C_2H_2. Reprinted with permission from Ref. 6. Copyright 1998, American Institute of Physics.

alternation of 3:1 ($J'' = $ odd : $J'' = $ even) due to nuclear spin statistics. The spectrum of the (2031^10^0) band is typical of a $\Pi_u - \Pi_g$ transition and consists of an unresolved Q-branch in addition to the P- and R-branches. The upper trace of Figure 7.3 shows a simulated vibrational spectrum, based on the rotational constants,[41–44] the nuclear spin degeneracies and the Boltzmann distribution at a rotational temperature of 80 K. The C_2H_2 rotational temperature was estimated from the characteristic rotational contour obtained in the action spectrum. Due to the presence of the hot-band transition $[(2031^10^0) \leftarrow (0001^10^0)]$ in the action spectrum, it was assumed that the vibrational temperature was higher than the rotational and that no vibrational relaxation occurred, *i.e.*, $T_{vib} = 300$ K.

As can obviously be seen, the intensities of the combination bands are far more prominent in the action than in the simulated spectra. The measured integrated intensity of the $(2030^00^0) \leftarrow (0000^00^0)$ transition is only about twice that of $(1410^02^0) \leftarrow (0000^00^0)$ and even less for the other band. This contrasts with the behavior encountered in absorption spectroscopy, where only the C–H stretch overtone transition was observed (due to poor sensitivity),[4–6] or was much more intense than the combination bands.[41–44] These observations manifest the difference between vibrational absorption and action spectra. Relying on the absorption spectra,[42–44] it was estimated that the $[(2031^10^0) \leftarrow (0001^10^0)] : [(2030^00^0) \leftarrow (0000^00^0)]$ intensity ratio was given by the $[(0001^10^0) : (0000^00^0)]$ population ratio, e.g., $\sim 4.5\%$ at 300 K. The absorption intensity of $[(1410^02^0) \leftarrow (0000^00^0)]$ relative to $[(2030^00^0) \leftarrow (0000^00^0)]$ was even lower and roughly estimated as $\sim 1\%$.

The dramatic difference in intensity ratios between the bands in the absorption and action spectra demonstrates their character. As shown in Section 2.1.2.2, in rovibrational absorption spectra the band intensities depend on the transition strength to the prepared vibrational state, while the action spectra depend, in addition, on the transition probability from the prepared state to the electronically excited states and possibly on the channels for photodissociation. Therefore, the enhanced action spectra of the (1410^02^0) and (2031^10^0) bands, which include bends, is related to a preferred transition probability from the prepared state to the upper bent electronic states, $\tilde{B}$ or $\tilde{A}$. From the above-mentioned experimental results, the roughly estimated enhancement factors are 50 and 16 for excitation from the (1410^02^0) and (2031^10^0) states, relative to that from the (2030^00^0) state, to the upper electronic states. Hence, energizing combination bands that include bending and particularly *cis*-bend resulted in vibrational wavefunctions on the ground state that improve the overlap with the dissociative wavefunction and drives the photodissociation more effectively than from a state containing mainly C–H stretch. This is supported by the *ab initio* calculation[32,34] that indicates the occurrence of adiabatic isomerization between the *trans* (more stable) structure and *cis*-isomer, from which the photodissociation takes place. Thus, bending excitation of the C≡C-H moiety contributes to the enhancement of the FC factor for the $\tilde{B}\,^1B_u/\tilde{A}\,^1A_u \leftarrow \tilde{X}\,^1\Sigma_g^+$ excitation that dissociates the molecule.

From the translational energy of the H photofragments, calculated from their Doppler profiles, it was concluded that the photodissociation of the different bands that mainly exhibit a regular rotational contour, occurs mostly through the $C_2H(\tilde{A}) + H$ channel. However, the $C_2H(\tilde{X}) + H$ channel, due to intersystem crossing or IC, could not be ruled out.[6]

(c) Rotational Effects in the $5\nu_{CH}$ Region. As can be noticed in Figure 7.3 from the comparison of the action spectra to the simulated spectra, the positions of the rotational lines in both spectra are the same, whereas the intensities are not. Most of the rotational transitions to the (1410^02^0) and (2030^00^0) states follow the intensities of the rotational contours depicted in the simulated spectra, representing a regular behavior, but the $R(13)$ and $R(23)$ transitions to the (2030^00^0) state, clearly marked in Figure 7.4, and several transitions to the (2031^10^0) state are more prominent than in the

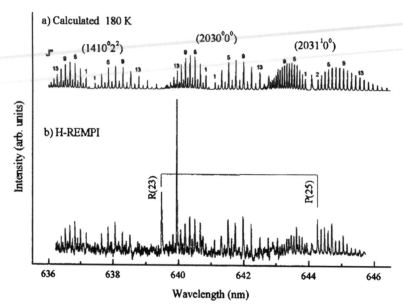

Figure 7.4 Vibrational overtone excitation spectra of the C_2H_2 $(1410^02^0) \leftarrow$ (0000^00^0), $(2030^00^0) \leftarrow (0000^00^0)$ and $(2031^10^0) \leftarrow (0001^10^0)$ transitions. The $R(13)$ line is the strongest in the H-REMPI spectrum. The prominent $R(23)$ and $P(25)$ lines that share the same upper J' are marked. The simulated absorption spectrum a) was calculated using rovibrational parameters from Refs. 41–44, nuclear spin degeneracies and the Boltzmann distribution *at a rotational temperature of 180 K*. The action spectrum b) presents the $(2+1)$ REMPI intensities of the H photofragments owing to 243.135-nm photolysis as a function of the visible wavelength applied for vibrational excitation. As in Figure 7.3, the simulated absorption intensities of the (1410^02^0) and (2031^10^0) bands are scaled up $\times 50$ and $\times 16$ relative to that of $((2030^00^0))$, respectively. The numbers above the peaks in the simulated spectrum assign the rotational levels of the vibrational ground state C_2H_2. Reprinted with permission from Ref. 6. Copyright 1998, American Institute of Physics.

simulated spectra. Moreover, as shown in Figure 7.4, when the action spectra were monitored at a higher rotational temperature, *i.e.* 180 K,[4-6] the intensities of the above-mentioned rotational transitions as well as that of $P(25)$ of the (2030^00^0) band were greatly enhanced. This behavior points to two kinds of irregularities: in the first one, both $R(23)$ and its counterpart $P(25)$, terminating at the same upper level ($J' = 24$), are enhanced; in the second one only $R(13)$, and *not* its $P(15)$ counterpart that terminates at the same upper level ($J' = 14$), shows anomalous high intensity. In addition, it was noticed[4-6] that the yield of the H atom in photodissociation *via* both $R(13)$ and $R(3)$ (a regular transition) shows a linear intensity dependence on the VIS excitation energy, with a two times larger slope for the former. The results indicate single-photon absorption in the VIS stage, but higher overall absorption cross section for $R(13)$ than for regular transitions. Moreover, for the $R(13)$ transition the Doppler profile is broader and the fragmentation pattern in the TOF spectrum is different, with a more pronounced C_2^+ signal than for regular transitions.[4-6] These observations are interpreted below.

Two different types of accidental double-coincidence resonances, capable of enhancing the two types of anomalously intense lines, were ascribed for the irregular rotations.[4-6] The suggested mechanism for the $R(23)$ and $P(25)$ transitions that show anomalous intensity and share the same upper J' is that of coupling between specific rotational levels of the upper vibrational state and energetically close-lying zero-order states. Presumably, these states (near $J'=24$) are dark zero-order states that carry no oscillator strength from the ground state, but enhance the signal in the electronic excitation step due to better FC factors. The involved dark states are assumed to contain bends and C≡C stretches possessing a better FC overlap with the bent upper state. This agrees with the above discussion about the large variation in signal intensity when preparing combination bands *vs.* C–H stretch overtone and also with the observed H Doppler profile and the fragmentation pattern.[4-6]

As for the "lonely" $R(13)$ transition, excitation of an additional weak overlapping transition, by the VIS photon, was proposed.[4-6] This transition (like others in C_2H_2[42-49]) is too weak to be observed in absorption, but shows up strongly *via* the subsequent UV electronic excitation and hence is observed in the action spectrum. The weak VIS transition is, possibly, to the $(0509^1 1^{-1})$ state.[42-44,47-49] This state contains nine quanta of *trans*-bend excitation and therefore expected to increase the FC factor for the UV excitation and to lead to higher intensity in the action spectrum. Since the VIS excitation of acetylene is in the $15\,600\ cm^{-1}$ region to a level containing so many bending quanta, the possibility of mixing-in vinylidene (isomer of acetylene) could also be suggested.[4-6] The weak VIS transition is followed, probably, by two UV photons that could be absorbed through an accidental double-coincidence resonance and differ from that of regular transitions where only one UV photon is absorbed. Due to the two-UV-photon absorption a sequential elimination of two H atoms or a synchronous C–H bond cleavage was inferred, favoring C_2^+ generation. This assumption is supported by the high intensity of the C_2^+ fragment in the TOF spectrum and by the broader H Doppler profile.[4-6]

(d) Comparison of VMP for Different Vibrational Levels in the $4\nu_{CH}$ Region.

Comparison of VMP for excitation of different vibrational levels was carried out also in the $4\nu_{CH}$ region of C_2H_2,[10] at somewhat higher total energy than the experiment reported above for $\nu_1 + 3\nu_3$ excitation.[2] Rovibrational excitation combined with ~ 243.1-nm promotion of C_2H_2 to the upper electronic *trans*-bent states, $\tilde{A}/\tilde{B}$, and H photofragment ionization-generated action spectra measuring the H yield as a function of the excitation wavelength. Similar to the above results for the $5\nu_{CH}$ region, the experiment in the $4\nu_{CH}$ region showed that the $(1030^0 0^0)$ IR bright state of the third C–H stretch overtone has a smaller photodissociation cross section than the $(1214^0 0^0)$ combination band. The latter contains *trans*-bend mode excitation, which enhances the FC factor for its excitation. Certain rotational line pairs, $R(J'-1)$ and $P(J'+1)$, originating from the (0000^00^0) state and accessing the same (high) J' levels of the (1030^00^0) state, showed anomalous intensities in the action spectrum. Also, the P transitions were excessively enhanced over the R transitions and, at least in one case, splitting of a P line was observed. The enhanced dissociation efficacy was interpreted as a consequence of Coriolis-type local resonances due, in particular, to overlap of the P transitions with those of nearby transitions of another band.[10]

(e) The $1\nu_{CH}$ Region.   To conclude the description of the studies on adiabatic VMP of C_2H_2 we allude to the study of the 197–212 nm photodissociation of acetylene from the ν_3 stretch at a combined energy of up to $53\,500\,cm^{-1}$.[11] From the kinetic energy of the H photofragment it was concluded that, as in other experiments, when photodissociation is carried out above the barrier for adiabatic process, the main dissociation channel is $C_2H(\tilde{A}) + H$.

7.1.2.2   Nonadiabatic VMP of C_2H_2

Nonadiabatic VMP of C_2H_2 is anticipated when the combined photodissociation energy is below the $\sim 52\,260\,cm^{-1}$ barrier for adiabatic dissociation to $C_2H(\tilde{A}\,^2\Pi) + H(^2S)$. VMP experiments that fulfill this condition were carried out for C_2H_2 prepared in the $3\nu_{CH}$ region and photodissociated by ~ 243-nm photons, the combined energy being $\sim 50\,800\,cm^{-1}$.[9,12,13]

For this energy, bound states on the $C_2H_2(\tilde{A}\,^1A_u)$ PES were reached, rather than the repulsive region (see Figure 7.5) and it was possible to monitor rovibronic levels of the $\tilde{A}$ state. To this end three lasers were applied, one for vibrational excitation of $C_2H_2(\tilde{X}\,^1\Sigma_g)$, one for electronic excitation and one for monitoring the H photofragments by $(2+1)$ REMPI (the excitations by these lasers are designated as NIR, UV_1 and UV_2, respectively, in Figure 7.5). The application of three lasers enabled UV action spectra to be obtained by scanning UV_1 across $\tilde{A} \leftarrow \tilde{X}$ rovibronic transitions, while parking the NIR and UV_2 wavelengths on a specific rovibrational transition of C_2H_2 and on the REMPI transition of H, respectively. NIR action spectra were also studied, using two lasers: NIR for scanning across the rovibrational transitions and UV at 243.135 nm for both $\tilde{A} \leftarrow \tilde{X}$ excitation and H atom REMPI. In addition, H Doppler profiles were monitored by fixing the NIR and UV_1 photolysis lasers on specific wavelengths and tuning the UV_2 probe across the H REMPI transition.

UV action spectra obtained by excitation of the initially prepared $J'' = 0$ of (1112^00^0) and (0030^00^0) to the $\tilde{A}\,^1A_u$ state of C_2H_2 are depicted in Figure 7.6. The wave number scale in the figure signifies the combined energy $(NIR + UV_1)$ for excitation of the corresponding rovibronic states of $\tilde{A}$. In order to compare the H^+ yield from the same rovibronic states, the UV_1 wavelength for exciting (1112^00^0) is shifted by $28.27\,cm^{-1}$ to the red relative to that for (0030^00^0). The frequencies of the sharp lines in both action spectra coincide, although they are more intense for (1112^00^0) where, in addition, more lines are observed. There

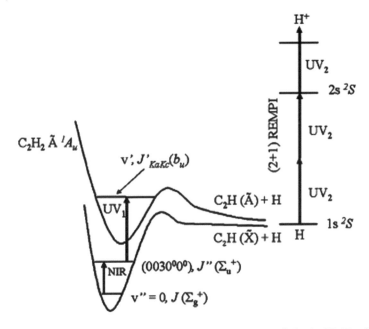

Figure 7.5 Schematics of the energetics relevant to nonadiabatic VMP of C_2H_2. Rovibrational excitation of C_2H_2 in the region of three C–H stretch quanta by NIR photons is followed by their promotion to the excited $\tilde{A}$ ^1A_u electronic state by UV_1 photons, where the molecules decomposed to release H photofragments that are detected *via* $(2+1)$ REMPI with UV_2 photons. Note that here J'' assigns a rotational state in the mediating vibrational level and J' in the $\tilde{A}$ ^1A_u state. Reproduced from Ref. 12 by permission of the PCCP Owner Societies.

are two possible explanations for this intensity difference. First, excitation to the (1112^00^0) and (0030^00^0) states leads to different FC factors for absorption of the UV_1 photons from these states. Second, the UV_1 photons populate rovibronic states in $\tilde{A}$ ^1A_u that couple well to dark states (see below) effectively promoting predissociation and hence H atom photofragmentation.

The three-laser experiment offers the possibility to discern between the two explanations since excitation from the two initially prepared rovibrational states leads to similar rovibronic states on the upper $\tilde{A}$ ^1A_u electronic state by using the same combined excitation energy. Accessing identical rovibronic states implies similar coupling to close-lying dark states and therefore the difference in H photofragment yield is a result of the different FC factors for the UV_1 absorption. Comparison of the two panels of Figure 7.6 indicates that the transition probabilities from $J''=0$ of (1112^00^0) are higher than those from $J''=0$ of (0030^00^0). The ratio of the integrated intensities of the transitions from (1112^00^0) to those from (0030^00^0) was found to be 10–68, with an average ratio of 27. This ratio accounts for the $\sim9\!:\!1$ intensity ratio of the $(0030^00^0)\!:\!(1112^00^0)$ measured *via* photoacoustic absorption spectroscopy.[12] Thus, energizing the (1112^00^0) combination band of acetylene, consisting of bending, leads to a

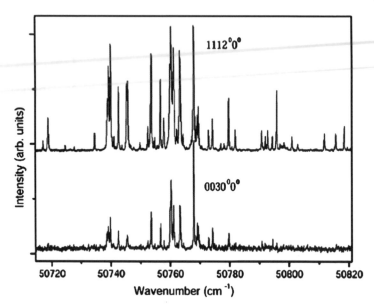

Figure 7.6 UV action spectra obtained by excitation of the initially prepared $J'' = 0$ of (1112^00^0) (upper panel) and of (0030^00^0) (lower panel) to the $\tilde{A}\,^1A_\mathrm{u}$ state of C_2H_2. The wave numbers signify the combined energy (NIR + UV$_1$) for excitation of the corresponding rovibronic states of $\tilde{A}\,^1A_\mathrm{u}$. Note that to get the same combined energy, the UV$_1$ wavelength for exciting (1112^00^0) is shifted by 28.27 cm^{-1} to the red relative to that for (0030^00^0). Reproduced from Ref. 12 by permission of the PCCP Owner Societies.

higher efficacy in dissociation. This is ascribed to the resemblance of the prepared vibrational wavefunction on the ground state to that on the upper state and accordingly to the improved FC overlap between them. Therefore, the enhanced FC factor for the intermediate state containing CH bending increases the $\tilde{A}\,^1A_\mathrm{u} - \tilde{X}\,^1\Sigma_g^+$ absorption and promotes the C–H bond cleavage more efficiently. The NIR action spectra supported the above conclusions:[12] Initial preparation of rovibrations of (1112^00^0) enhanced the H photofragment production by about an order of magnitude over that of the preparation of (0030^00^0).

The results of the three-laser experiments (Figure 7.6) are consistent with nonadiabatic VMP since they show that bound states of $\tilde{A}\,^1A_\mathrm{u}$ are reached, where adiabatic dissociation is not accessible. The nonadiabatic nature of the dissociation was corroborated by the translational energies of the H photofragments obtained from their Doppler profiles.[12] It was found that in the VMP from different rovibrational states of C_2H_2 the departing H atoms were released with a maximal translational energy of 0.55 ± 0.07 eV. The translational energy is close to the 0.58 eV difference between the overall photon energy acquired by the molecule *via* the NIR + UV excitation (6.29 eV) and the energy required for photodissociation through the nonadiabatic channel $C_2H(\tilde{X}) + H$

(5.71 eV).[32,34] This indicates that the VMP of C_2H_2 occurs nonadiabatically under the present conditions.

7.1.3 VMP of C_2HD

7.1.3.1 Adiabatic VMP of C_2HD – Evidence for Rotational Dependence of H/D Branching Ratio

Similarly to HOD, VMP of C_2HD offers two distinct isotopic channels. To study these channels, C_2HD was excited, alternatively, to five stretches of CH ($5v_1$), four stretches of CH and one of CD ($4v_1 + v_3$), or three stretches of CH plus one of CC, one of CD and two *cis*-bends ($3v_1 + v_2 + v_3 + 2v_5$). These vibrational states were promoted to the $\tilde{A}\,^1A_u$ and $\tilde{B}\,^1B_u$ states by 243.135- and 243.069-nm photons that were also used for monitoring, respectively, the H and D photofragments, *via* (2 + 1) REMPI.[3,7,8] Figures 7.7 and 7.8 represent action spectra of H and D photofragments (panels b and c) as a function of rovibrational excitation and the simulated absorption spectrum (panel a) of the relevant bands. The $H + C_2D$ and $D + C_2H$ photofragments were the outcome of C–H and C–D bond cleavage, respectively. The simulations were based on known molecular constants of the $5v_1$ and $4v_1 + v_3$ states.[19,20] The simulations represent spectra of $\Sigma^+ - \Sigma^+$ vibrational bands including *P*- and *R*-branches and that of $5v_1$ is similar to the Fourier transform vibrational spectrum.[19] The H and D peaks (panels b and c) track the positions of the rotational lines in the simulation (panel a). The yield of both H and D photofragments was enhanced upon photolysis with that of H being higher than that of D. From the Doppler profiles of both H and D it was concluded that the ~ 243.1 dissociation of C_2HD in the $5v_1$ region is adiabatic.

Although the excitation coefficient of the $5v_1$ state is $< 10^{-6}$ of the fundamental C–H stretch,[20] a large signal is observed in the action spectra, showing extremely efficient production of H and D. The intensity ratio between $5v_1$ and $4v_1 + v_3$ as measured from the amplitudes of regular *R* transitions (those that do not stick out of the spectrum) ($J' = 6$–8) was 1.5 ± 0.1. This ratio agrees well with that of 1.55, measured from absolute transition dipole moments in direct-absorption experiments.[19,20] The almost comparable intensity ratios in action spectra and in absorption indicate similar character of the two bands that contain only stretches. This is probably due to the fact that in the second step of excitation to the upper electronic states the transitions do not differ much in their FC factors. As for $3v_1 + v_2 + v_3 + 2v_5$, no data concerning its intensity in direct absorption is available and therefore the influence of the FC factor on the action spectrum intensity could not be determined.

As in C_2H_2, the most intriguing observation is that pairs of rotational transitions terminating in the same upper J' ($J' = 2, 11, 12, 17$–$19, 21, 22$ of $5v_1$ and $J' = 1, 9, 12, 15$–$17, 22$ of $4v_1 + v_3$), in both the H and D action spectra, are more pronounced and stick out from the *P* and *R* regular rotational contour. It is seen that some of the transitions exhibit an intensity of more than an order of magnitude higher than that for a Boltzmann distribution. In contrast, the

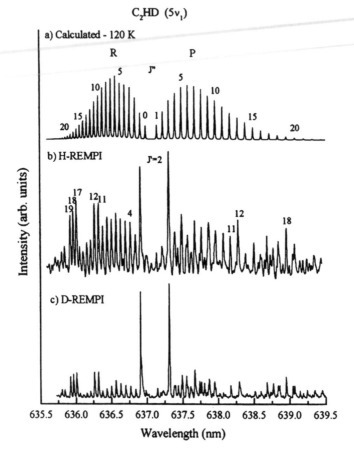

Figure 7.7 Vibrational overtone excitation spectra of the $5v_1$ state of C_2HD. (a) simulated absorption spectrum calculated using rovibrational parameters from Ref. 19 at a temperature of 120 K. Panels (b) and (c) show the action spectra of H and D photofragments as a result of ~ 243.1 nm photolysis. The numbers above the peaks in the simulated spectrum label the rotational levels of ground-state C_2HD and those in the action spectrum mark the rotational levels of $5v_1$ for which preferential enhancement is observed. Reprinted from Ref. 7. Copyright 1999, with permission from Elsevier.

absorption spectrum of $5v_1$ varies smoothly with J' and follows a Boltzmann distribution.[19,20] The observation that both *P*- and *R*-branch transitions sharing similar J' deviate from the regular behavior was attributed, as in C_2H_2 [$R(23)$ and $P(25)$], to couplings with dark zero-order states.

Moreover, the rovibrational preparation did not affect only the transition intensities in the action spectra, but also the H/D branching ratio. For example, for $5v_1$, the H/D branching ratios were 4.9 ± 0.5 for the regular line (*e.g.*, $J' = 4$) and 1.5 ± 0.2 for the irregular (*e.g.*, $J' = 2$).[3,7] For $4v_1 + v_3$, the H/D ratio for

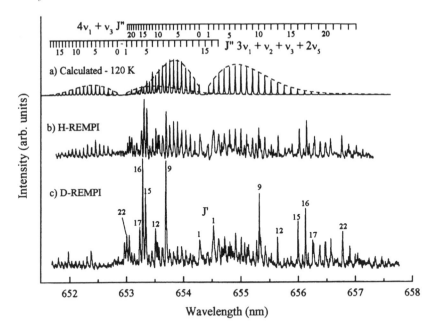

Figure 7.8 Vibrational overtone excitation spectra of the $4v_1 + v_3$ and $3v_1 + v_2 + v_3 + 2v_5$ states of C_2HD. (a) Simulated absorption spectrum calculated using rovibrational parameters from Ref. 20 at 120 K. (b) and (c) action spectra of H and D photofragments as a result of ~ 243.1-nm photolysis. The rulers above the simulated spectrum label the rotational levels of ground-state C_2HD and those in the action spectrum mark the rotational levels of $4v_1 + v_3$ that are standing out from the rotational contour. Adapted from Ref. 7 with permission from Elsevier.

regular lines was in the range of 2.5–3.8, while for irregular transitions it was 0.9–1.5.

The results for the regular lines demonstrate that there is a change in the yield and in the H/D branching ratio following bond fission in C_2HD ($5v_1$) and C_2HD ($4v_1 + v_3$), also as a result of preparation of different vibrational states. The combined energy (VIS + UV) decreases only slightly from ~ 7.05 eV for $5v_1$ to ~ 7.0 eV for $4v_1 + v_3$ and therefore it is unlikely to be the reason for the difference in branching ratio. A possible explanation is that in $5v_1$ only the C–H bond is extended, while in the $4v_1 + v_3$ state the C–D bond is extended as well. In both cases the C–H bond is cleaved preferentially, but the H/D ratio of 4.9 for regular lines in the former is reduced to 2.5–3.8 in the latter. The effect is even more dramatic when irregular rotational states are prepared prior to photodissociation since then the preference is reduced to almost statistical H/D ratios (0.9–1.5). This is, most probably, a result of the coupling of the prepared J' with zero-order dark states. Possible candidates for interaction are states that include combinations of C–D stretches and bends or C≡C stretches and bends. This interaction leads to extension of the C–D bond and to enhancement of its cleavage as a result of better FC overlap, which is represented by the higher intensities in the action spectra and, moreover, by the different H/D branching ratios.

The observation that the same pairs of P and R transitions, terminating at the same J's, are more pronounced in the action spectra of both H and D, points, indeed, to resonances of these J's with other states that are then more effective in excitation to the upper electronic state. *To conclude, the rotational specific VMP of* C_2HD *serves as a fascinating example of product identity and H/ D branching ratio control not only via excitation of different skeletal motions, but even by preparation of neighboring rotational states.*

7.1.3.2 Nonadiabatic VMP of C_2HD

The nonadiabatic VMP of C_2HD[14] was carried out under similar conditions and using similar techniques as that for C_2H_2 (Section 7.1.2.2). However, C_2HD was prepared in the (2100^02^0) and (3000^00^0) states (rather than (1112^00^0) and (0030^00^0) in C_2H_2) and both the H and D photofragments were monitored. As for C_2H_2, it was found that the initial vibrational preparation alters the FC factor for the subsequent transition to the upper $\tilde{A}$ state, resulting in preferential enhancement of the C–H(D) bond cleavage for excitation from the (2100^02^0) state that contains bends over that from (3000^00^0) that consists of stretches only. From the H and D UV action spectra and from the Doppler profiles it was concluded that, not surprisingly, the ~ 243.1 dissociation of C_2HD in the $3\nu_1$ region is nonadiabatic. The H/D branching ratio for photodissociation from the (2100^02^0) state was found to be 1.3 ± 0.3.

7.2 VMP of Ammonia Isotopologues

7.2.1 Spectral Features of Ammonia Relevant to VMP

The photodissociation of ammonia at energies relevant to VMP takes place *via* excitation to the $(\tilde{A}\,^1A''_2)$ state:[50–54]

$$NH_3(\tilde{X}\,^1A'_1) + h\nu \rightarrow NH_3(\tilde{A}\,^1A''_2) \rightarrow NH_2(\tilde{X}\,^2B_1) + H(^2S) \qquad (7.3)$$

$$\rightarrow NH_2(\tilde{A}\,^2A_1) + H(^2S) \qquad (7.4)$$

where similar schemes apply to the other isotopologues. The VMP scheme for NH_3 is described in Figure 7.9. The ground state of NH_3, $\tilde{X}\,^1A'_1$, has pyramidal structure, whereas the first electronically excited state, $\tilde{A}\,^1A''_2$, is planar (trigonal).[50] For excitation energies sufficient for dissociation *via* both channels, the branching ratio between the nonadiabatic channel (eqn (7.3)) and the adiabatic channel (eqn (7.4)) is governed by the nonadiabatic transition that takes place in the vicinity of the CI located around $r(N–H) = 1.5$ Å.[50–54]

VMP studies of ammonia isotopologues were conducted on NH_3,[55–61] NHD_2[62] and NH_2D.[62] The four (six if one counts degeneracies) fundamental

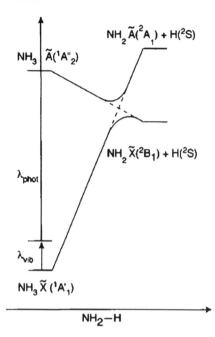

Figure 7.9 Excitation scheme for VMP of NH_3. The dotted curves indicate the CI between the two electronic states. The photodissociation is monitored, alternatively, *via* emission from $NH_2(\tilde{A}\,^2A_1)$, $(2+1)$ or $(3+1)$ REMPI of $H(^2S)$. Reprinted with permission from Ref. 56. Copyright 2002, American Institute of Physics.

vibrational levels of the ground electronic state of NH_3 are[15] the symmetric $v_1 = 3337\,\text{cm}^{-1}$ stretch, the $v_2 = 950\,\text{cm}^{-1}$ umbrella (inversion), the doubly degenerate antisymmetric $v_3 = 3444\,\text{cm}^{-1}$ stretch and the doubly degenerate $v_4 = 1627\,\text{cm}^{-1}$ bend. VMP of NH_3 was carried out from each (or from combinations) of the fundamental vibrations[55–57,59–61] and from the $4v_1 + v_3$ and $5v_1$ states.[58] For the partly deuterated isotopologues, VMP was performed on $5v_{NH}$ of NHD_2 and $5v_{ND}$ of NH_2D, where the respective frequencies are[63] $5v_{NH} = 15\,524\,\text{cm}^{-1}$ $5v_{ND} = 11\,734\,\text{cm}^{-1}$.

7.2.2 VMP of NH_3

7.2.2.1 VMP of NH_3 Prepared at Low Vibrational Energies

It might look surprising that one of the first VMP experiments where a specific vibrational level of a molecule was prepared prior to its excitation to a dissociative state was carried out on NH_3.[55] This is probably due to the fact that at the time that this experiment was done, CO_2 lasers were popular (and practically the only far-IR lasers) and their wavelengths around $\sim 10.6\,\mu m$ coincide with those needed to excite the v_2 level of NH_3. Although the photodissociation products were not monitored (the purpose of this study was to measure vibrational population *via* changes in electronic absorption), the combined energies of the photons in the experiments (at $10.6\,\mu m$ plus 200–225 nm) were above the dissociation threshold. It was shown that it is possible to excite 40% of the NH_3 molecules by a pulsed CO_2 laser under the given experimental conditions.

However, in later VMP experiments on NH_3, state-to-state processes were investigated. In this subsection and in 7.2.2.2 we present the studies on mode-specific dependence of the photofragment yields and in 7.2.2.3 the application of VMP for spectroscopic studies of NH_3.

A decisive experiment on the mode specificity of the photofragment yields is presented in Ref. 56, where NH_3 was prepared, alternatively, in the v_1, v_3, $2v_4$, $v_1 + v_2$, $v_2 + v_3$ and the $v_2 + 2v_4$ state. The vibrationally excited NH_3 was promoted to the $\tilde{A}\,^1A''_2$ state using a *tunable* UV laser in order to get *isoenergetic photolysis* of the NH_3, the specific dissociation energy chosen for this study being $52\,878\,cm^{-1}$. The photodissociation was monitored *via* emission from the $NH_2(\tilde{A}\,^2A_1)$ photofragment. Since this is a clear example where preparations of different vibrational states, followed by dissociation at exactly the same total energy, lead to different distributions of products, we dwell on this example in some detail.

The room-temperature photoacoustic absorption spectra, the vibrational action spectra at $T_{rot} \sim 10\,K$ and simulations of the absorption spectra in the N–H stretching region from 3150 to $3500\,cm^{-1}$ (v_1, v_3 and $2v_4$) and from 4100 to $4580\,cm^{-1}$ ($v_1 + v_2$, $v_2 + v_3$ and $v_2 + 2v_4$) are depicted in Figures 7.10 and 7.11, respectively. Note that each of the rotational levels of the symmetric top ground-state NH_3 splits into a doublet (due to inversion): a symmetric sublevel (s, positive vibrational parity) and an antisymmetric (a, negative vibrational parity) with respect to reflection in the inversion plane.[64] The notation used in Figure 7.10 is $s/a^{\Delta K}\Delta J(J'',K'')$ where ΔK or $\Delta J = -1,0,1$ are denoted by P, Q, R, respectively.

The results of the experiments and simulations presented in Figures 7.10 and 7.11 are summarized in Figure 7.12. The upper panel in the figure compares the intensities in the action spectrum for v_1, v_3 and $2v_4$ (solid bars) to those in the simulation absorption spectrum, both normalized to the largest band in the simulation (v_1), and the lower panel compares these intensities for $v_1 + v_2$, $v_2 + v_3$ and $v_2 + 2v_4$. Whereas the difference between the measurement and simulation in the upper panel is striking, there is hardly any difference in the lower panel. In the former, both the antisymmetric stretching band (v_3) and the bending overtone band ($2v_4$) are three times more intense in the vibrational action spectrum than the simulation predicts. In sharp contrast, in the latter the relative yield is insensitive to initial vibrational excitation of the combination bands that contain the umbrella (inversion) motion v_2.

As explained in Section 2.1.2.2, vibrational absorption spectra depend on the transition probability for the vibrational transition, whereas vibrational action spectra depend, in addition, on the transition probability for photodissociation out of the prepared intermediate state to form the monitored dissociation product. The vibrational mode dependence of the $NH_2(\tilde{A}\,^2A_1)$ yield could arise from a better FC factor for electronic excitation, depending on both lower and

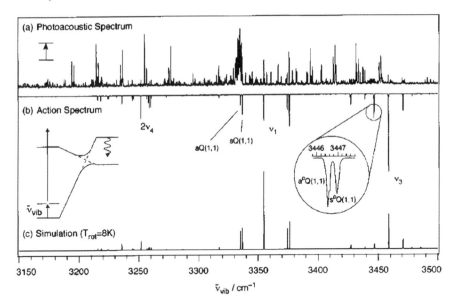

Figure 7.10 (a) Room-temperature photoacoustic absorption spectrum, (b) jet-cooled action spectrum and (c) simulation of the absorption spectrum for $T_{rot} = 8\,K$, of NH_3 in the N–H stretching region. The inset shows an example of the doublet corresponding to the two components of the inversion splitting. Reprinted with permission from Ref. 56. Copyright 2002, American Institute of Physics.

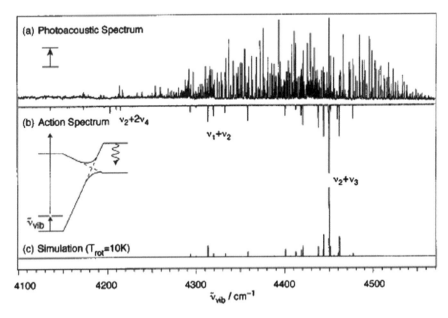

Figure 7.11 (a) Room-temperature photoacoustic absorption spectrum, (b) jet-cooled action spectrum and (c) simulation of the absorption spectrum for $T_{rot} = 10\,K$, of NH_3 in the N–H stretch-umbrella combination band region. Reprinted with permission from Ref. 56. Copyright 2002, American Institute of Physics.

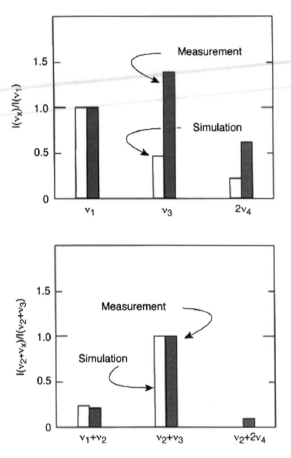

Figure 7.12 Integrated band intensities of the six vibrational bands of NH₃ in the action spectrum (solid bars) compared to the simulation, revealing (upper panel) more efficient production of electronically excited fragments for both the antisymmetric stretch (v_3) and the bending overtone ($2v_4$). The additional quantum of umbrella motion (lower panel) removes any distinction among the initial excitations in the N-H stretch-umbrella combination band region. Reprinted with permission from Ref. 56. Copyright 2002, American Institute of Physics.

upper state vibrational wavefunctions or from an increase the probability of a nonadiabatic transition to form the ground-state $NH_2(\tilde{X}\,^2B_1)$ product and thus *decrease* the $NH_2(\tilde{A}\,^2A_1)$ product yield. The vibrational action spectra presented above cannot distinguish between the two mechanisms.

Further, detailed insight into mode specificity in the VMP of NH₃ was obtained when the H photofragments were detected by $(2+1)^{60}$ or $(3+1)^{61}$ REMPI and their Doppler profiles[60] or VMI monitored.[61] Both methods enable the kinetic energy of the products to be obtained, but the latter enables measurement of the recoil energy with a much higher resolution and determination of the internal energy of the unobserved NH₂ product. The most dramatic observation in these studies was the difference between the outcome of dissociation from the symmetric stretch, designated below as 1^1, *vs.* that from the antisymmetric stretch, 3^1, of $NH_3(\tilde{A}\ ^1A''_2)$. These dissociative states were prepared, respectively, by IR excitation of a single rovibrational symmetric stretch, 1_1 or antisymmetric stretch, 3_1, in $NH_3(\tilde{X}\ ^1A'_1)$, followed by UV excitation. The combined IR + UV excitation was the same in both cases. An illuminating demonstration of this difference is depicted in Figure 7.13. The following conclusions were gathered from the measurements presented by the figure:[60] The Doppler profile for dissociation from 1^1 is much broader than that from 3^1. The average translational energy release for dissociation from 1^1 is $3490 \pm 450\,cm^{-1}$, 10 times more than that for dissociation from 3^1. The speed distributions are also very different with a bimodal distribution for 1^1 with maxima at 5520 and 13 850 m/s and a single maximum for 3^1 corresponding to slow hydrogen atoms with speeds < 5000 m/s. This implies that dissociation from 1^1 is mainly through the nonadiabatic channel, with some contribution from the adiabatic channel, whereas dissociation from 3^1 is almost solely *via* the adiabatic channel.

In order to further interpret the above differences we refer to Figure 7.14, where calculated $\tilde{X}$- and $\tilde{A}$-state PES of NH_3 along one N–H bond distance $R(NH_2–H)$ and the out-of-plane bending angle θ are shown. As can be seen in the figure, the surfaces form a CI, crossing in the planar geometry ($\theta = 90°$). Detailed interpretation is presented in Ref. 61, where a primary goal was to

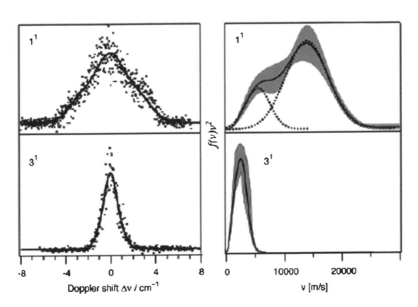

Figure 7.13 Doppler profiles and speed distributions for dissociation from the symmetric stretch (upper panel) and antisymmetric stretch (lower panel) of $NH_3(\tilde{A}\ ^1A''_2)$. The dotted lines in the speed distribution for 1^1 show both slow and fast components, and the shaded area is the standard deviation. Reprinted with permission from Ref. 101. Copyright 2003 American Chemical Society.

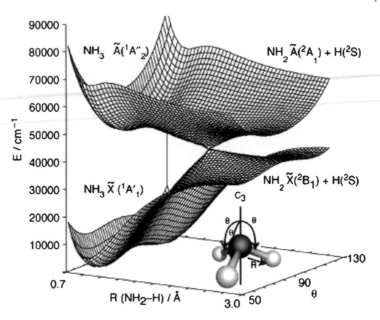

Figure 7.14   Calculated $\tilde{X}$- and $\tilde{A}$-state PESs (PES) of NH_3 along one N–H bond distance $R(NH_2-H)$ and the out-of-plane bending angle θ. The surfaces form a CI, crossing in the planar geometry ($\theta = 90°$). Reprinted with permission from Ref. 61. Copyright 2006, American Institute of Physics.

compare VMP *via* the 1^1 and 3^1 state channels, applying VMI in order to identify the electronic state of the NH_2 fragment for each cannel. Using this technique, the extent of adiabatic and nonadiabatic dissociation for different vibrations in electronically excited NH_3 is unraveled. The images of the VMI show that decomposition of NH_3 with the 1^1 excited symmetric N–H stretch produces primarily ground-state NH_2 fragments, $NH_2(\tilde{X}\ ^2B_1)$ (the nonadiabatic channel). By contrast, decomposition from electronically excited NH_3 containing one quantum of the 3^1 antisymmetric N–H stretch leads almost exclusively to electronically excited fragments $NH_2(\tilde{A}\ ^2A_1)$ (the adiabatic channel), avoiding the CI between the excited-state and ground-state surfaces. In other words, the production of $NH_2(\tilde{A}\ ^2A_1)$ suggests that the antisymmetric stretching excitation in the electronically excited NH_3 carries it away from the CI that other vibrational states access. It is worth noting that it was shown that there are additional important CIs between the ground and excited states of NH_3.[65] Time-dependent quantum wavepacket calculations that take these findings into account would help explain the fine details of the photodissociation dynamics of ammonia.[61]

7.2.2.2   *VMP of NH_3 Prepared at High Vibrational Energies*

Measurements and calculations of VMP from high vibrational states of NH_3, aiming to compare the cross sections of N–H bond dissociation in the photolysis from the $4v_1 + v_3$ and $5v_1$ states, are reported in Ref. 58. The experimental procedure combined measurements of the absorption spectrum of NH_3 for vibrational excitation into the $4v_1 + v_3$ and $5v_1$ states, applying PA spectroscopy, and of the action spectrum resulting from dissociation from these state, *via* $(2+1)$ REMPI of the H photofragments. In the computational procedure the FC factors for absorption from the vibrational states were evaluated by utilizing time-dependent wavepacket calculations on the PESs constructed by an *ab initio* molecular-orbital procedure.

It was found that the cross section for dissociation from $4v_1 + v_3$ is larger than that from $5v$ by a factor of 1.23 ± 0.06, whereas the theoretical ratio was calculated as 1.02. The discrepancy between the experimental and theoretical results was attributed to two nonexclusive possible reasons. (1) Not taking into account in the calculations the vibrational interactions between the stretching and bending modes. (2) Alterations (caused by the different vibrational excitations) of the measured quantum yield for the N–H dissociation channels (eqns (7.3) and (7.4)) due to the (not measured) $NH + H_2$ channel.[58]

7.2.2.3 *VMP of NH_3 as a Spectroscopic Tool*

In addition to providing insight into the photodissociation dynamics, the VMP of NH_3 provided unique information on the vibrational structure of its $\tilde{X}$ and $\tilde{A}$ states.[56,57] In the VMP study described above (Section 7.2.2.1), where the photodissociation was monitored *via* emission from the $NH_2(\tilde{A}\ ^2A_1)$ photofragment, a total of 60 individual rotation–inversion transitions for six vibrational bands of $NH_3(\tilde{X})$ were assigned or confirmed.[56]

Information on the vibronic structure of the $\tilde{A}$ state and in addition on the lifetimes of the vibronic states was obtained from the action spectra, monitored *via* $(2+1)$ REMPI of the H photofragments of jet-cooled NH_3 initially excited to selected rovibrational states.[57] We quote below some of the highlights of this study. As in many other cases described throughout this monograph, it was also found here that initial vibrational excitation significantly changes the FC factors for the subsequent electronic transition. This was instrumental for the assignment of sharp resonances to a progression in the degenerate bending mode including the $2^1 4^1$ and $2^3 4^1$ combination states (the superscripts are the number of quanta in each state) that were not assigned before. Also identified were broad, non-Lorentzian shaped resonances as transitions to the symmetric and antisymmetric stretching fundamentals. The proposed assignment revealed an accidental degeneracy of the umbrella (v_2) and bending mode (v_4) with fundamental frequencies of $892 \pm 8\ cm^{-1}$ and $906 \pm 30\ cm^{-1}$, respectively.

Simulation of the vibronic band envelopes provides band origins and homogeneous rovibronic linewidths for states containing the umbrella and bending modes.[57] The homogeneous linewidths Γ were converted to lifetimes (see Section 2.2.2), $\tau = (2\pi c \Gamma)^{-1}$, which are plotted as a function of excess energy over that of the 0^0 state in Figure 7.15. The lifetimes of the bending states decrease monotonically with excess energy from 115 ± 18 fs for 4^1 to 13 ± 4 fs for 4^4, with lifetimes at least a factor of 2 less than for nearly

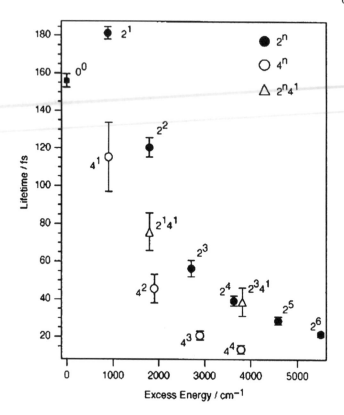

Figure 7.15 NH$_3$($\tilde{A}$) lifetimes extracted from the homogeneous rovibronic linewidths Γ using the relation $\tau = (2\pi c \Gamma)^{-1}$ plotted as a function of excess energy over that of the 0^0 state. Reprinted with permission from Ref. 57. Copyright 2002, American Institute of Physics.

isoenergetic excitation of the umbrella motion. The combination states $2^1 4^1$ and $2^3 4^1$ on the other hand, live longer than the pure bending states but shorter than states involving only the umbrella motion.[57] Since there were discrepancies between the presently observed and previously calculated lifetimes, *ab initio* recalculations of the barrier in the exit channel of NH$_3$($\tilde{A}$) were carried out.[57] These calculations yielded a barrier to dissociation of 2348 cm^{-1}, about 20% lower than previous calculations and partly explained the discrepancy between measured and calculated lifetimes, the latter being longer. For a better fit, improvement of the PES of NH$_3$($\tilde{A}$) is needed.

7.2.3 VMP of NHD$_2$ and NH$_2$D

VMP of NHD$_2$ and NH$_2$D is yet another test of vibrationally mediated bond selectivity where preferential production of either H or D atoms may be monitored. Such preferential production was encountered in triatomic molecules, HOD (Section 6.4), and linear tetratomic molecules where a rigid

C≡C bond separates between the H and D contain moieties (Section 7.1.3). The nonlinear, tetratomic NHD_2 and NH_2D molecules present more complex systems where vibrationally mediated H and D production was studied.[62]

In the work described in Ref. 62 the preparation of NHD_2 in the $5\nu_{NH}$ vibrational state and NH_2D in the $5\nu_{ND}$ state was followed by ~ 243.1 nm excitation to the dissociative $\tilde{A}$ state. The H and D photofragments were monitored (by the same UV laser used for excitation) *via* (2 + 1) REMPI. Figure 7.16 depicts the REMPI signals for fixed wavelengths for preparation of specific rovibrational states and scanning of the UV wavelengths to cover the isotope shift between H- and D-atom REMPI spectra. The H/D yield ratio was found by comparing the areas in Figure 7.16. Considering the number of NH and ND bonds in NHD_2 and NH_2D, the ratio of the NH dissociation cross section to that of ND was found to be 5.1 ± 1.4 and 0.68 ± 0.16 for NHD_2 ($5\nu_{NH}$) and NH_2D ($5\nu_{ND}$), respectively. These results were compared with those for the photolysis of vibrationally unexcited NH_2D and NHD_2.[66,67] It was concluded that the $5\nu_{NH}$ excitation in NHD_2 leads to enhanced NH dissociation with a ~ 2 times larger NH/ND branching ratio than for the vibrational

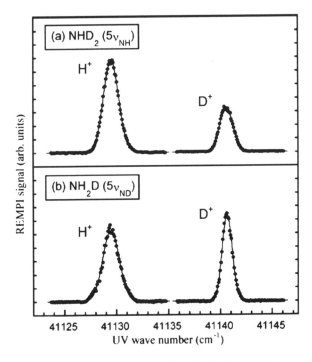

Figure 7.16 The (2 + 1) H and D REMPI signals for the VMP of (a) $NHD_2(5\nu_{NH})$ and (b) $NH_2D(5\nu_{ND})$ as a function of the UV-laser wave number. Reprinted with permission from Ref. 62. Copyright 2004, American Institute of Physics.

ground state of NHD_2; the $5\nu_{ND}$ excitation in NH_2D leads to preferential ND dissociation with a ~ 3–4 times larger ND/NH branching ratio than for the vibrational ground state of NH_2D.

Interpretation of the experimental results was attempted considering the energetics and the wavepacket dynamics.[62] Time-dependent wavepacket calculations, similar to those applied for NH_3 $4\nu_1 + \nu_3$ and $5\nu_1$ (Ref. 58 and Section 7.2.2.2), were performed utilizing the PES constructed by an *ab initio* molecular orbital calculation. The results of the calculations showed that the NH (ND) stretching-excitation leads to the nearly complete selection of the NH (ND) bond dissociation. This is in contrast to the bond selectivity observed in the experiment that was substantially lower. The discrepancy was attributed to the fact that the effect of the bending modes was not considered in the wavepacket calculations and to nonadiabatic crossings between PESs. It was pointed out that exact wavepacket calculations in full-dimensional surfaces (that might be very difficult) should provide information about the effect of the different vibrational modes and on the contribution of the nonadiabatic dynamics.[62]

7.3   VMP of HNCO

7.3.1   Spectral Features of HNCO Relevant to VMP

VMP of isocyanic acid, HNCO,[68–83] is a heuristic case study, despite its complex spectroscopy, since it tests vibrational state control of photodissociation into *three* distinct channels, two of them leading to chemically distinct species:[83]

$$HNCO(\tilde{X}\,^1A') + h\nu \rightarrow HNCO(\tilde{A}\,^1A'')$$

$$\rightarrow NH(X\,^3\Sigma^-) + CO(X\,^1\Sigma^+) \quad D_0 = 30\,080\,\text{cm}^{-1} \quad (7.5)$$

$$\rightarrow H(^2S) + NCO(\tilde{X}\,^2\Pi) \quad D_0 = 38\,320\,\text{cm}^{-1} \quad (7.6)$$

$$\rightarrow NH(a\,^1\Delta) + CO(X\,^1\Sigma^+) \quad D_0 = 42\,710\,\text{cm}^{-1}. \quad (7.7)$$

We refer hereafter to $NH(X\,^3\Sigma^-)$ as ^3NH and to $NH(a\,^1\Delta)$ as ^1NH. The ground state of HNCO is planar, nonlinear, nearly an accidental prolate symmetric top, where the NCO moiety is nearly linear with the NCO angle of $173°$ and the HNC angle of $124°$.[84] The PES of the excited $(\tilde{A}\,^1A'')$ state is presented in Figure 2.1, where a two-dimensional cut through the six-dimensional PES is depicted as a function of the two bond distances R_{HN} and R_{NC}. The two exit channels on this PES, (7.6) and (7.7), are presented in Figure 2.1, channel (7.5), which involves intersystem crossing to the a triplet state, is not shown in the figure but indicated in Figure 7.17.

The six vibrational NMs of $HNCO(\tilde{X})$ consist of three stretches and three bends. The fundamantal streches are[76,85] the N–H stretch, $\nu_1 = 3538\,\text{cm}^{-1}$, and the antisymmetric and symmetric stretches of the N–C–O moiety, $\nu_2 = 2269\,\text{cm}^{-1}$

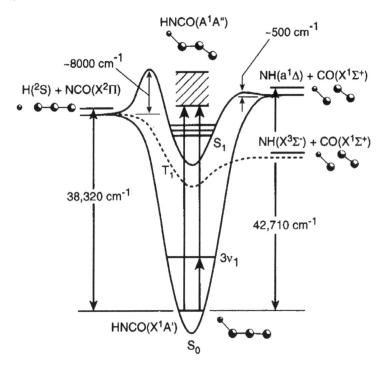

Figure 7.17 Schematics of one-photon dissociation and of VMP of HNCO prepared in the $3v_1$ state. The minimum-energy curves for HNCO show the energies of the products, the origin of the S_1 state and the calculated barriers to dissociation on S_1. The hatched region indicates the energies accessed in the one-photon and in the VMP experiments described in Section 7.3.3. Reprinted with permission from Ref. 83. Copyright 2001, American Institute of Physics.

and $v_3 = 1323\,\text{cm}^{-1}$, respectively. The fundamental bends are[76,86] the H–N–C bend, $v_4 = 777\,\text{cm}^{-1}$, the N–C–O bend, $v_5 = 577\,\text{cm}^{-1}$ and the torsion, $v_6 = 656\,\text{cm}^{-1}$.

The extensive VMP studies of HNCO included preparation of fundamental vibrations, overtones and combination states. In the following sections we first (Section 7.3.2) explain what could be anticipated in these studies and briefly survey the experimental methods applied and the results obtained in VMP of HNCO prepared in various vibrational states. Then (Section 7.3.3) we describe the detailed study of VMP of HNCO, prepared in the $3v_1$ state, in comparison to one-photon dissociation of thermal molecules. We conclude with the application of VMP to HNCO spectroscopy (Section 7.3.4).

7.3.2 A Brief Survey of the VMP of HNCO

Figure 7.17 is a schematic illustration of a specific example, VMP of HNCO prepared in the $3v_1$ state, and of one-photon dissociation at the same energy.[83]

Examining the figure the following observations could be made. First, the dissociation at energies smaller than $\sim 45\,000\,\mathrm{cm}^{-1}$ may involve at least three electronic states, two singlet states, S_0 and S_1, and one triplet state, T_1. Secondly, below the threshold for spin-allowed N–C bond cleavage, where only the $^3\mathrm{NH} + \mathrm{CO}$ and $\mathrm{H} + \mathrm{NCO}$ channels are open, competition between those two channels is expected; the spin-forbidden $^3\mathrm{NH} + \mathrm{CO}$ channel requires intersystem crossing where HNCO has to reach a triplet state such as T_1. Thirdly, above the threshold, where formation of $^1\mathrm{NH} + \mathrm{CO}$ is also possible, competition among all three channels is expected. Fourthly, even when the excitation energy exceeds the threshold for production of $\mathrm{H} + \mathrm{NCO}$, the large barrier in that channel prevents direct dissociation on S_1 and may bring the ground electronic state S_0 into play. Finally, since in the electronically excited state of HNCO the N–C–O moiety is bent, whereas in the ground state it is nearly linear, it is expected that bending excitation might enhance the FC factor for electronic transition.

Selective vibrational preparation of HNCO was carried out by either VIS/IR excitation[68–70,77,78,81,83] or SRE.[71,73,75,76,80] The VMP products, NH and NCO, were detected by LIF.[68–71,73,75–78,80,81,83] In the first VMP study of HNCO, dissociation from the $4v_1$ state, prepared *via* excitation at 759 nm, was compared to single-photon dissociation at the same total energy, about $1000\,\mathrm{cm}^{-1}$ above the threshold for the $^1\mathrm{NH}$ channel, and found to enhance the efficiency of the NCO channel over the $^1\mathrm{NH}$ channel by a factor of at least 20. It was estimated that the $\mathrm{NCO}/^1\mathrm{NH}$ branching ratio in the one-photon dissociation at this energy is ≈ 20, and it grows to ≥ 400 in the vibrationally mediated photodissociation.[68] The next VMP study of HNCO was out of the $3v_1$ state; we deal in detail with this state in Section 7.3.3. Some of the other VMP studies of HNCO are related to its spectroscopy and are alluded to in Section 7.3.4; however, we mention here some conclusions that are instrumental for the understanding of the VMP dynamics.

The effect on VMP of mixing the fundamental vibrational stretches of HNCO (v_1, v_2 and v_3) with other vibrational states was systematically investigated.[73,75,76] The v_1, v_2 and v_3 states were prepared by SRE and the $^1\mathrm{NH}$ and NCO products detected by LIF. Analyzing the LIF action spectra, it was found that preparation of unperturbed v_1 (N–H stretch) states leads to diffuse NCO yield spectra compared to excitation of mixed vibrational states.[75] The higher-energy dissociation channel that produces $^1\mathrm{NH}$ has coarser structure near its threshold, consistent with a more rapid dissociation. When SRE preparation of v_1 was followed by comparing the intensities of the PARS and action spectra, the effect of coupling v_1 to the other vibrational states could be unraveled.[73] Extensive state mixing of v_1 was observed and possible v_1 perturbers, consisting of various combinations of the v_2, v_3, v_4, v_5 and v_6 states, were suggested. Particularly relevant to VMP is the fact that the intensity differences between PARS and action spectrum transitions, which revealed the vibrational state mixing, provided FC factors for transitions to the dissociative electronic state. Similar comparison between the intensities of the PARS and action spectra, where the v_2 and v_3 states were prepared by SRE, unraveled the couplings of these states.[76] It was found that v_3 is mixed with a combination containing two quanta of bends, the strongest perturber being (probably) $v_5 + v_6$ and (possibly) a weaker one being $v_4 + v_5$. Although v_2 was also found to be strongly perturbed, the smaller Raman cross section precluded detailed spectroscopic exploration. However, for both v_3 and v_2 the influence of the initial preparation and mixing on the FC factors was evaluated. It was concluded that promotion of out-of-plane bend decreases the FC factor since the electronic transition in hand is between two states with planar equilibrium geometries.

In order to identify intramolecular couplings that control the early stages of IVR following vibrational preparation of HNCO, room-temperature PA spectra of $2v_1$ to $5v_1$ and free-jet action spectra of ^{1}NH, ^{3}NH and NCO following pre-excitation of $3v_1$ to $5v_1$ of the N–H stretching were studied.[78] Analyzing the spectra, it was concluded that the states most strongly coupled to the pure N–H stretching zero-order states are ones with a quantum of N–H stretching excitation v_1 replaced by different combinations of N–C–O asymmetric or symmetric stretching excitation (v_2 or v_3) and the v_4 bending excitation. The two strongest couplings of the nv_1 state are to the states $(n - 1)$ $v_1 + v_2 + v_4$ and $(n - 1)v_1 + v_3 + 2v_4$, and sequential couplings through a series of low-order resonances potentially play a role. The analysis also showed that if the pure N–H stretch zero-order state were excited, energy would initially flow out of that mode into the strongly coupled mode in 100 fs to 700 fs, depending on the level of initial excitation. At this point it is worth mentioning the one-photon studies of HNCO, where it was suggested that there is a contribution to the photolysis signal from vibrational hot bands, in particular from the lowest vibrationally excited state, the 577 cm^{-1} bend (v_5), which has a population of about 6% compared to the ground state in the 300 K experiment.[87] It was pointed out that due to the geometry change upon excitation, hot-band transitions from an excited bending state are likely to have favorable FC factors and contribute significantly to the photolysis signal.

The effect of coupling to the bending state was studied *via* absorption and action spectroscopy, and theoretical calculations, on states in the $3v_1$ region of HNCO.[81] Couplings between the bright N–H stretching state and states that have one or more quanta of N–C–O bending excitation were identified. The mixed state is much more efficiently photodissociated than the pure N–H stretching zero-order state. Because the N–H bond length in the S_1 state of HNCO is similar to that for the ground state, excitation of the N–H bond in the ground state does not improve the FC factor for the $S_1 \leftarrow S_0$ transition. However, because N–C–O, which is linear in the ground state, is bent in the electronically excited state, N–C–O bending excitation significantly improves the excitation probability.

7.3.3 Comparison of VMP of HNCO $3v_1$ to One-Photon Dissociation of Thermal Molecules

In this section we closely follow, in some detail, Ref. 83, where previous and additional studies, comparing VMP of HNCO $3v_1$ to one-photon dissociation, are critically analyzed and summarized. Reference to Figure 7.17 above may serve as a visual aid. The comparison was carried out *via* direct observation of all three photofragmentation channels of HNCO and measurements of the relative yields of $^3NH + CO$, $H + NCO$, and $^1NH + CO$ at nine different photolysis energies for both thermal and vibrationally excited molecules. Figure 7.18 depicts the LIF excitation spectra of the NH and NCO products of

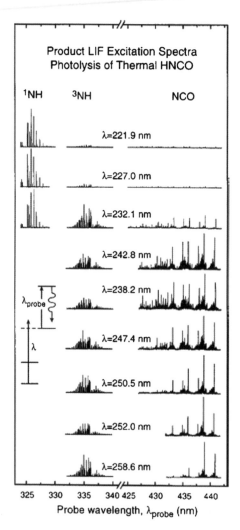

Figure 7.18   LIF excitation spectra of the NH and NCO products of the one-photon photodissociation of thermal HNCO molecules. The lower six spectra are for photolysis energies where $^3NH + CO$ and $H + NCO$ are the only possible products. The three spectra at the top are for photolysis energies where production of $^1NH + CO$ is possible as well and show the dominance of the ^1NH product channel in those energy regimes. Reprinted with permission from Ref. 83. Copyright 2001, American Institute of Physics.

the one-photon photodissociation of thermal HNCO molecules at the different wavelengths. The figure clearly shows that as the photolysis energy increases, the LIF excitation spectrum of NCO extends over a larger range of wavelengths, indicating increasing excitation of the bending vibration of the NCO fragments with increasing available energy. However, as the three top spectra in the figure indicate, when the photolysis energy exceeds the threshold for the spin-allowed cleavage of the N–C bond to form ^{1}NH + CO, the LIF signal from ^{1}NH grows sharply with increasing energy, reflecting the dominance of this dissociation pathway once it is energetically accessible.

Figure 7.19 compares the LIF excitation spectra of the NH and NCO products from the one-photon dissociation of thermal HNCO to that from vibrationally excited HNCO at the same total excitation energies, one below the threshold for ^{1}NH production (lower pair of spectra) and one above the threshold (upper pair of spectra). At the lower total energy, only the ^{3}NH + CO and the H + NCO channels are open. The notable result is that the yield of ^{3}NH relative to that of NCO does *not* change between the photolysis of vibrationally state-selected molecules and thermal molecules. At the higher total energy, where all three channels are open, vibrational excitation changes the relative yields markedly. As in the lower-energy photolysis, the yield of ^{3}NH relative to NCO does not change with vibrational excitation, but the relative yield of NCO and ^{3}NH compared to ^{1}NH increases substantially in the photolysis of vibrationally excited HNCO.

The analysis of the spectra presented in Figures 7.18 and 7.19 is displayed in Figure 7.20. It quantitatively summarizes the results presented in the previous figures and adds the results of some other studies of one-photon dissociation of HNCO.[88–90] Although some of the details presented in the figure are mentioned above, we emphasize below the conclusions that can be drawn from the figure, due to the unique feature of the VMP of HNCO, namely, the vibrational state control of photodissociation into *three* distinct channels.

The solid bars in Figure 7.20, representing the quantum yields for the one-photon dissociation of thermal HNCO, show that the yield of ^{3}NH + CO decreases and that of H + NCO rises at the threshold for forming the latter products. Formation of NCO becomes the dominant channel as the photolysis energy increases until the energy is enough to form ^{1}NH + CO, which then become the dominant, but not sole, products. When both spin-allowed channels are open, about 80% of the products are ^{1}NH + CO and about 20% are H + NCO up to at least 3000 cm^{-1} above the threshold for ^{1}NH production. The LIF excitation spectra in Figure 7.18 and the corresponding relative yields shown in Figure 7.20 demonstrate the dominance of each successive channel with increasing energy. The open bars in Figure 7.20, representing the quantum yields for the VMP of HNCO($3\nu_1$), show that at energies below the threshold for production of ^{1}NH + CO, the relative yields of ^{3}NH and of NCO are the same for the photolysis of vibrationally state-selected or thermal HNCO molecules. The significant difference appears above the threshold for ^{1}NH + CO formation. Comparison of the heights of the open bars (VMP) and the closed bars (one-photon photolysis) shows that photodissociation of HNCO($3\nu_1$)

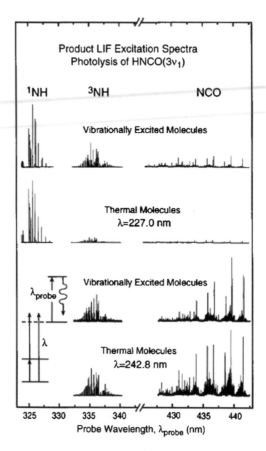

Figure 7.19   LIF excitation spectra of the NH and NCO products of the photodissociation of thermal HNCO molecules and of vibrationally excited
HNCO molecules. The lower two spectra are for a total photolysis
energy where $^3NH + CO$ and $H + NCO$ are the only possible products.
The top two spectra are for energies where production of $^1NH + CO$ is
possible as well. In both cases, the photolysis wavelength used for the
vibrationally excited molecules produces the same total energy content as
in the one-photon photolysis. Reprinted with permission from Ref. 83.
Copyright 2001, American Institute of Physics.

yields substantially more NCO compared to ^1NH than does the photodissociation of thermal molecules.

   To get more insight into the complex one-photon and VMP decomposition
of HNCO, a five-dimensional classical trajectory calculation of the photodissociation of HNCO in the S_1 electronic state was carried out.[79] It suggested
that the mechanism of vibrational enhancement is reduction of the direct dissociation to $^1NH + CO$ on S_1. It was found that trajectories on S_1 for HNCO
molecules with two or four quanta of N–H stretching excitation evolve to
products more slowly than those at the same energy with no initial vibrational
excitation. The average lifetime of the molecules increases by about a factor of

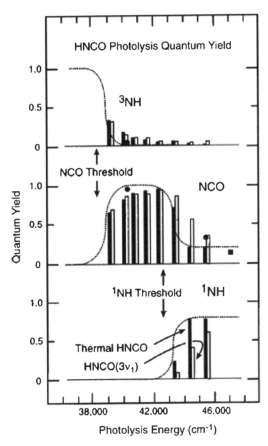

Figure 7.20 Quantum yields of ^{3}NH (top panel), NCO (middle panel), and ^{1}NH (bottom panel). The solid bars are the quantum yields for the one-photon dissociation of thermal HNCO, and the open bars are the quantum yields for the VMP of HNCO($3v_1$). The circles are the results for the 248- (Ref. 88) and 222- (Ref. 89), and the square for the 212-nm one-photon photolysis (Ref. 90). The dotted lines are smooth curves drawn through the one-photon photolysis data. Reprinted with permission from Ref. 83. Copyright 2001, American Institute of Physics.

5. Initial motion along the N–H coordinate is orthogonal to the N–C direction along which dissociation to ^{1}NH + CO occurs and the molecule does not move along the dissociation direction as readily. The increased dissociation lifetime allows the molecule to sample the crossing region to S_0 more times and, hence, increases the probability of forming NCO and ^{3}NH *via* the ground electronic state.

To conclude, measurements and calculations of the relative yields of each of the three channels, ^{3}NH + CO, H + NCO, and ^{1}NH + CO, in the photolysis of both thermal and vibrationally excited HNCO molecules provided information on the competition among the adiabatic and nonadiabatic (spin-forbidden) pathways. The dependence of the yields on vibrational excitation supports a picture in which direct decomposition on the S_1 surface produces $^1NH + CO$ and in which IC to S_0 leads to $H + NCO$, by spin-allowed unimolecular decay, and to $^3NH + CO$ by intersystem crossing and decomposition on T_1. More specifically, for dissociation below the ^1NH threshold S_0 is the state that is common to production of both NCO and ^3NH with the former products arising from the unimolecular decomposition on S_0 and the latter from subsequent intersystem crossing from S_0 to T_1. Once the system reaches S_0, competition between statistical N–H bond fission to form $H + NCO$ and the intersystem crossing to T_1 determines the relative yield. While initial vibrational excitation of HNCO might influence the rate of IC from S_1 to S_0, the subsequent dynamics on S_0 are likely to erase any memory of the initial vibrational excitation and make the relative yields of NCO and ^3NH independent of the vibrational excitation of HNCO. The indifference of the relative yield to HNCO vibrational excitation is strong experimental evidence for the system reaching T_1 by first crossing to S_0.

Although vibrational excitation does not alter the relative amounts of ^3NH and NCO it has a marked effect on the relative amount of ^1NH. Photolysis of HNCO $(3v_1)$ yields about 2.5 times *less* ^1NH than does photolysis of a thermal sample with the same total energy. This is consistent with the initial N–H stretching excitation impeding the direct decomposition on S_1 to form $^1NH + CO$, as suggested by the trajectory calculation of Ref. 79, allowing a greater opportunity for IC to S_0 and subsequent decomposition into the competing $H + NCO$ and $^3NH + CO$ channels. *In other words, the control in VMP of HNCO comes from vibrational excitation hindering one pathway and enhancing the others.* It is worth noting that relying on the results of the spectroscopic study of HNCO presented in Ref. 80 (see below), it was suggested that the N–C stretching vibration is a promoting mode for IC from S_1 to S_0.

7.3.4 Application of VMP to HNCO Spectroscopy

Some of the above-mentioned studies deal with or allude to spectral features of HNCO obtained *via* its VMP studies, in particular Refs. 71,73,75,76,78,80 and 81. The conclusions drawn from the spectroscopic measurements, which are directly related to our main preoccupation, VMP dynamics, were presented in the previous sections. Here, we present some highlights of HNCO spectroscopy (or spectroscopy related) obtained *via* VMP.

We first note the measurements of the upper limits for the bond enthalpies of the N–H and C–N bonds in HNCO by observation of photodissociation appearance thresholds for the NCO($\tilde{X}\ ^2\Pi$) and NH($a\ ^1\Delta$) fragments from initially selected HNCO vibrational states.[71] The complex experimental setup consisted of four separate light sources: one fixed-frequency and one tunable source for the SRE preparation of the vibrationally excited state (v_1), a tunable photolysis laser for the electronic excitation, and a tunable probe laser to detect either NCO($\tilde{X}\ ^2\Pi$) or NH($a\ ^1\Delta$) fragments by LIF. The upper limit of the dissociation energy of the H–N bond was found to be $D_0(\text{H–NCO}) \leq 38$ $320 \pm 140\,\text{cm}^{-1}$ and that of the N–C bond $D_0(\text{HN–CO}) \leq 42\,710 \pm 100\,\text{cm}^{-1}$.

The VMP-Raman[73,75,76] and the VMP-absorption spectroscopy[78,81] studies mentioned above (Section 7.3.2) provided information on spectroscopic features of rotational and vibrational levels of the ground electronic state. We dwell here on the vibrational structure of the S_1 state, studied *via* mediated photofragment yield spectroscopy combined with multiphoton fluorescence spectroscopy and detailed simulation.[80] The experimental approach taken in this study is illustrated in Figure 7.21 and described in the caption.

The vibrationally mediated photofragment yield spectroscopy relies on SRE preparation of vibrational states in S_0 that have good FC factors for subsequent excitation to low-lying vibrational levels in S_1 that, in turn, radiationlessly decays to the triplet state T_1 (shown but not mentioned in the figure) that decomposes to NH $(X\,^3\Sigma^-) + \text{CO}(X\,^1\Sigma^+)$. The electronic spectrum in the region of the origin is obtained by scanning the wavelength λ_{elec} of the electronic excitation laser while monitoring the NH $(X\,^3\Sigma^-)$ decomposition product by LIF. In the multiphoton fluorescence spectroscopy, one photon with a wavelength λ_{elec} excites a transition to a low-lying vibronic state in S_1, and a second photon from the same laser pulse carries the electronically excited molecule to a higher-lying, dissociative electronic state in which HNCO decomposes to form $\text{H}(^2S) + \text{NCO}(A\,^2\Sigma^+)$. Monitoring the emission from the electronically excited NCO while varying the laser wavelength yields an electronic excitation spectrum of HNCO molecules cooled in a supersonic expansion. The mediated photofragment yield and the multiphoton fluorescence spectroscopies are complementary. This is due, mainly, to the fact that the SRE applied for vibrational state preparation of HNCO has the advantage of improving the FC factor for transitions to states near the origin. It has the disadvantage of exciting Q-branch transitions that lead to rotational congestion. On the other hand, the approach using multiphoton excitation has the advantage of exciting molecules from a cooled sample in which there is little rotational congestion.

From the analysis of the spectra the origin of the S_1 state is located at 32 $449 \pm 20\,\text{cm}^{-1}$ and the harmonic vibrational frequencies of the N–C stretch $(\omega_3 = 1034 \pm 20\,\text{cm}^{-1})$, the H–N–C bend $(\omega_4 = 1192 \pm 19\,\text{cm}^{-1})$ and the N–C–O bend $(\omega_5 = 599 \pm 7\,\text{cm}^{-1})$, are found. As mentioned at the end of the previous section, the assigned spectra suggest that the N–C stretching vibration is a promoting mode for IC from S_1 to S_0.

7.4 VMP of Other Tetratomic Molecules

In this section we describe VMP studies on hydrogen peroxide isotopologues (HOOH and HOOD),[91–102] nitrous acid (HONO)[103,104] and hydrazoic acid (HN$_3$).[105,106] Since mode-selective or bond-selective dissociation was not observed in experiments on these molecules, the description will be short as compared to that in the previous sections.

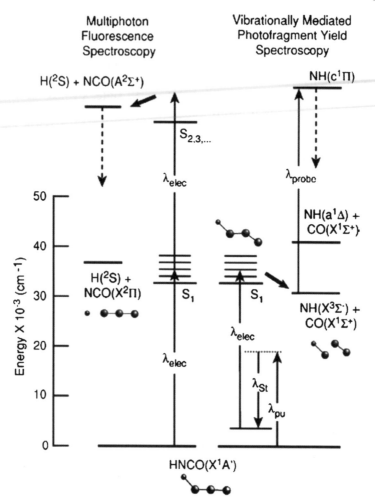

Figure 7.21   Energy-level diagram illustrating vibrationally mediated photofragment yield spectroscopy and multiphoton fluorescence spectroscopy of HNCO. The right-hand side shows vibrationally mediated photofragment yield spectroscopy in which SRE prepares a vibrational state (by the pump, λ_{pu}, and the Stokes, λ_{St}, photons), a photon (λ_{elec}) dissociates the vibrationally excited molecule, and LIF (following excitation at λ_{probe}) detects the triplet NH product. The left-hand side shows multiphoton fluorescence spectroscopy in which one photon excites HNCO to the S_1 state and another photon of the same wavelength promotes the molecule to a higher-energy, dissociative state, forming electronically excited NCO whose fluorescence is detected. Reprinted with permission from Ref. 80. Copyright 2000, American Institute of Physics.

7.4.1 Hydrogen Peroxide Isotopologues

The ground electronic state of HOOH is nonplanar with six normal vibrational modes:[100] $v_1 = 3599\,cm^{-1}$, OH symmetric stretch, $v_2 = 1387\,cm^{-1}$, symmetric bend, $v_3 = 875\,cm^{-1}$, OO stretch, $v_4 = 254\,cm^{-1}$, torsion, $v_5 = 3611\,cm^{-1}$, OH asymmetric stretch and $v_6 = 1265\,cm^{-1}$, asymmetric bend. VMP of HOOH[91–102] (including theoretical work[100]) and HOOD[97,99] was studied for species prepared by initial excitation in the region of the third ($4v_{OH}$) overtone of the OH stretching and in some cases also in the region of the fourth overtone ($5v_{OH}$).[92,94] The OH or OD photoproduct were detected *via* LIF.

In a series of experiments on HOOH, comparison of VMP to one-photon dissociation, including thermally assisted vibrational-overtone-induced predissociation through the $5v_{OH}$ transition, or to two-photon vibrational-overtone-induced predissociation was carried out at room temperature.[92,94] The energy-level diagram for the dissociation of HOOH related to these experiments is depicted in Figure 7.22. It was found that the photodissociation dynamics in VMP is different from that in one-photon UV dissociation of comparable or greater energy. Thus, in the VMP of HOOH ($4v_{OH}$), 11% of the OH fragments are formed in $v = 1$, in contrast to the one-photon UV dissociation where essentially all of the produces are formed in $v = 0$.

The experiments and interpretation presented in Refs. 92 and 94 were followed by modeling in Ref. 100 where the authors presented LM basis set calculations for the region of the HOOH ($4v_{OH}$) wavefunction. The results support the dissociation model outlined in Ref. 92. Furthermore, the calculated enhancement of the photodissociation cross section from the overtone state relative to that from the ground state agrees qualitatively with experiment, as does the increase in vibrationally excited OH products. The mechanism for overtone-mediated photodissociation relies on the overtone wavefunction having two key characteristics. First, it is highly localized in the overtone degree of freedom, *i.e.* it has a large overlap with the zero-order overtone and insures a large cross section for the initial excitation to the overtone. Second, the wavefunction has very delocalized tails that extend into the wide amplitude O–O stretch region. These tails, although small, play the important role of providing FC overlap with the continuum states, necessary for describing the dissociation mechanism.

In calculating the wavefunctions in the region of HOOH ($4v_{OH}$), a global PES was used and LM product basis functions that are prediagonalized in each mode so as to yield a very good zero-order description of the overtone wavefunction were constructed. The tails arise from mixing with nearly degenerate background states, which are themselves extensively mixed. This framework provided a novel view of the extent of this mixing by demonstrating the sensitivity of the overtone wavefunction to its precise location relative to the nearly degenerate background states. The dissociation cross section was calculated by evaluating FC overlaps of the overtone wavefunction with continuum states on a repulsive electronic surface. A separable approximation was used to calculate these overlaps; partially removing this approximation produced no qualitative

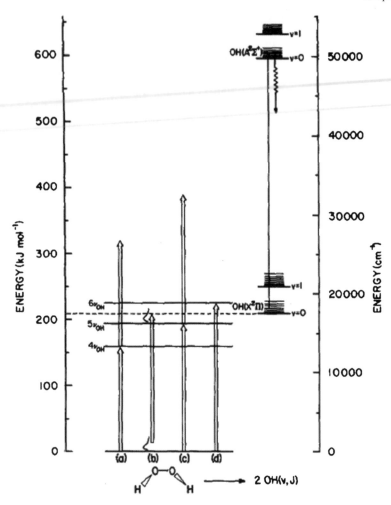

Figure 7.22   Energy-level diagram for the dissociation of HOOH. The dashed line is the thermodynamic O–O bond dissociation threshold. (a) VMP through the $4v_{OH}$ vibrational overtone transition. The first photon excites the $4v_{OH}$ vibrational transition and the second (with the same energy) excites the vibrationally excited molecules to a dissociative electronic state. (b) Thermally assisted vibrational overtone induced predissociation through the $5v_{OH}$ transition. In this one-photon process, molecules that possess sufficient initial thermal energy dissociate following excitation of the $5v_{OH}$ transition. (c) VMP through the $5v_{OH}$ vibrational overtone transition. The first photon excites the $5v_{OH}$ transition and the second promotes the vibrationally excited molecule to a dissociative electronic state. (d) Vibrational-overtone-induced predissociation through the $6v_{OH}$ transition. Reprinted with permission from Ref. 92. Copyright 1987, American Institute of Physics.

changes in the conclusions. In evaluating the continuum wavefunctions, an *ab initio* surface was used and model surfaces were constructed and used for comparison. For all surfaces it was found that the VMP cross section was enhanced by several orders of magnitude over the direct UV cross section at the same energy, and that the partial cross sections for vibrationally excited OH products were much greater than those for the direct transition. A significant aspect of these calculations was finding that the torsional mode plays a crucial role in the VMP process.[100]

The VMP of HOOH ($4v_{OH}$) was reinvestigated in a jet-cooled setup[101] and revealed anomalous double-resonance spectral intensities. The relative photodissociation cross sections out of the intermediate $4v_{OH}$ state were determined as a function of the J'_{KaKc} level, where Ka and Kc are the projections of the total angular momentum J on the a- and c-axes. HOOH ($4v_{OH}$) molecules with $J' = 0$ were found to have a negligible photodissociation cross section compared with levels with J' and $K'a > 0$. Combined with information about the OH(X) $v = 0$ rotational state distribution generated *via* photodissociation of HOOH ($4v_{OH}$, $J'_{KaKc} = 2_{02}$), the cross-section data suggest that delocalization of the intermediate overtone state wavefunction into wide-amplitude O–O stretching regions, which is required in the VMP experiment to provide good FC overlap with the excited electronic continuum, is profoundly influenced by parent molecular rotation, in particular about the a- and b-axes. The absence of VMP transitions involving the $4v_{OH}$ $J' = 0$ level implies that the intermediate state for $J' = 0$ is more highly localized than its $J' > 0$ counterparts. The last observation may be of significance in the context of the search of mode-selective chemistry.

Vector correlations,[93,96,97,99] vibrational and rotational fragment pair (scalar) correlations[97–99] and population distributions in OH photofragments generated *via* VMP of HOOH($4v_{OH}$)[93,96–99] and in OD photofragments generated *via* VMP of HOOD($4v_{OH}$)[97,99] at selected wavelengths near 750 nm, were determined using polarized, Doppler-resolved LIF spectroscopy. These measurements, together with a knowledge of the direct, *single*-photon dissociation dynamics at comparable supplied energies, allowed assignment of the two electronic transitions ($\tilde{A}^1A$ and $\tilde{B}^1B \leftarrow \tilde{X}^1A$) in hydrogen peroxide, contributing to the continuum accessed in the double-resonance experiments, and established the minor influence of rotational torques exerted on the electronically excited PESs, when these are accessed at large O–O bond extensions. In addition, the measurements established strong photofragment rotational excitation as originating primarily from the molecular rovibrational motion in the intermediate high-overtone state and allowed the use of an approximate classical treatment to estimate the average energy disposed into each vibrational mode in the vibrationally excited HOOH(D) molecules "interrogated" by secondary photon absorption at ~ 750 nm. The "interrogated" molecules, initially excited *via* the $4v_{O–H}$ "LM" transition, were found to possess significant OOH(D) bending and HO–OH(D) torsional character in addition to contributions from HO–OH(D), H–O, and D–O stretching. Also, it was found that the intramolecular vibrational mode composition together with the component of parent molecular rotation about the *a*-axis is effectively mapped into the final photofragment quantum-state distributions.

7.4.2 Nitrous Acid

The VMP of nitrous acid, HONO, was studied both theoretically[103] and experimentally.[104] A classical trajectory study of IVR, *cis–trans* isomerization and unimolecular dissociation in HONO are presented in Ref. 103. The calculations were carried out on a realistic PES that was constructed by using the available kinetic, thermochemical, spectroscopic, and *ab initio* quantum-mechanical information. The influence of the total energy, initial NM excitations, initial OH-stretch overtone excitations, rotations, and PES on IVR and the initial rates of isomerization and dissociation of HONO were analyzed. The results showed significant mode-specific behavior, particularly for the isomerization. It was shown that excitations of overtones of the O–H or N = O bond-stretching modes yield the lowest initial rates for both isomerization and dissociation. Excitation of the HON bending mode yields the largest isomerization rates, while excitation of the ONO bending mode yields the largest dissociation rate. At a fixed total energy, placing a small amount of rotational energy in the molecule causes a significant increase in the isomerization and dissociation initial rates over those computed for nonrotating HONO, however, when the rotational energy is increased above 0.1 eV, the rates decrease as expected on the basis of the Rice–Ramsperger–Kassel–Marcus (RRKM) theory. The orientation of the rotation is an important factor for the intramolecular energy transfer and reaction rates. Rotating the molecule along the torsional axis causes a significant increase of the initial rate of isomerization. Rotating the molecule perpendicular to the ON bond causes significant increases in the dissociation rates.[103]

The experimental investigation of the spectroscopy and dynamics of HONO encountered problems that are mentioned in Ref. 104 and it is worth repeating them here since the merit of VMP as a tool for solving these problems is heuristically exemplified. Both calculations and experiments suggest that the energy barrier to *cis–trans* isomerization by torsional motion is of the order of $3500\,cm^{-1}$ (*trans*-HONO has the thermodynamically preferred conformation: $\Delta E(cis–trans) = 141 \pm 36\,cm^{-1}$). There do not appear to have been any reliable estimates of the magnitude of the barrier to isomerization *via* in-plane bending, nor have there been many reports of any overtone or combination absorption bands of either *cis* or *trans* nitrous acid. This, in part at least, is due to the fact that it is not possible to obtain a pure sample of HONO. Rather, it will always be found as a (minor) constituent in a mixture containing NO, NO_2 and H_2O. Thus, the difficulties of working with a low species concentration is further compounded by problems of spectral overlap, which cannot easily be removed by rotational cooling given the reported ease with which HONO molecules cluster in a molecular beam. VMP offered a means of alleviating these problems since, as shown in previous chapters, overtone absorptions are detected by monitoring the yield of photofragments that arise as a result of a further photon absorption that projects the vibrationally excited molecules up to a (dissociative) excited electronic state. Specifically, spectra of the O–H stretching overtones of HONO, free of any overlap from absorptions associated with the various contaminants mentioned above, could be obtained by monitoring resonance enhancements in the excitation spectrum for forming OH(X) fragments.

Reference 104 describes LIF measurements of the energy disposal in the OH(X) photofragments resulting from VMP of the second ($3v_1$) and third ($4v_1$) O–H stretch overtones of *trans*-HONO. As in one (UV) photon study of HONO photodissociation from its first excited ($\tilde{A}$ $^1A''$) electronic state, these OH(X) fragments were observed to be formed exclusively in their zero-point vibrational level. However, these fragments (especially those formed in the VMP *via* the third O–H stretch overtone) showed significantly higher levels of rotational excitation than those found in the single UV-photon dissociations. These differences were used to provide some insight into the intramolecular dynamics within the vibrationally excited parent molecule.[104]

7.4.3 Hydrazoic Acid

The two published works on the VMP of hydrazoic acid, HN_3, dealt with the vector correlations following 532-nm photodissociation from the second overtone of the N–H stretching vibration ($3v_1$), monitoring the NH(a $^1\Delta$) photofragment by LIF.[105,106]

In the first study,[105] the energy disposal and the ν–J vector correlation in the nearly isoenergetic HN_3 one-photon 355-nm photodissociation and 532-nm VMP of the $3v_1$ level were compared. A significantly hotter rotational state distribution of the NH(a $^1\Delta$) fragment was observed in the latter process, indicative of altered dissociation dynamics. Comparison of Doppler profiles of P and Q lines for LIF detection of the NH(a $^1\Delta$) fragment in a given rotational level for the VMP process revealed that the NH fragments possessed a substantially greater ν–J correlation with both vectors tending to align themselves in a much more collinear fashion than in the nearly isoenergetic one-photon photodissociation. This means that although the two photodissociation experiments correspond to similar total energies, excitation from initial nuclear configuration with an extended N–H bond apparently generates much larger torsional forces, which tend to align the fragment's rotational angular momentum and recoil velocity vectors in a collinear fashion. It was suggested that the differences in the torsional forces experienced by the molecule in the two experiments result from variation in the region of the excited state surface accessed by them. Furthermore, since *ab initio* calculations[107] found only one singlet state ($\tilde{A}$ $^1A''$) in this wavelength region, the differences in the two isoenergetic measurements most likely arise from excitation of the parent molecule to different regions of the *same* electronic surface and not from accessing different excited electronic states in the two experiments.[105]

The second study[106] concentrated on detailed analysis and interpretation of Doppler profiles of specific rotational levels of the NH(a $^1\Delta$) fragment obtained following the 532-nm VMP of HN$_3$ ($3v_1$). To this end, a general expression for the Doppler profile for fragments produced in the photo-dissociation of laser-excited, aligned molecules, as in VMP, was derived by employing a semiclassical approach.[106] It was found that the values of parameters obtained in the analysis of Doppler profiles for the detection of NH rotational levels $N = 7$ and 10 are quite different from those predicted by simple expectations of dissociation through a perpendicular transition between planar geometries. Thus, the experimental data are consistent with a value of $\sim 0°$ for the angle θ_a between μ and the parent molecule a inertial axis, while a value of 90° is predicted for the $\tilde{A}$ $^1A'' \leftarrow \tilde{X}$ $^1A'$ transition, and the recoil anisotropy has a positive value of 0.2–0.3, in contrast to the predicted value of $-1/2$. Both of these results are strongly suggestive of the importance of nonplanar geometries, as has also been inferred for the one-photon photolysis of HN$_3$.[108]

The derived values for the parameters θ_a and the recoil anisotropy reveal different aspects of the dissociation dynamics.[106] The angle θ_a depends on the geometry of the molecule at the instant of electronic excitation. By contrast, the recoil anisotropy parameter depends both on the direction of μ in the molecular frame and on the dissociation dynamics following electronic excitation. A likely explanation for the unexpected derived value for θ_a could come from nonplanar geometries accessed directly in the electronic excitation. The total excitation energy in the 532-nm VMP of HN$_3$ ($3v_1$) is considerably less than the vertical $\tilde{A}$ $^1A'' \leftarrow \tilde{X}$ $^1A'$ excitation energy, $\sim 41\,000\,\mathrm{cm}^{-1}$.[107] It was found in Ref. 107 that the minimum excitation energy, $\sim 30\,000\,\mathrm{cm}^{-1}$, is to a nonplanar geometry with an out-of-plane N–N–N angle of $\sim 60°$. Because of the low excitation energy in the VMP process, $\sim 28\,500\,\mathrm{cm}^{-1}$, the excitation is not in the vertical FC regime, and nonplanar geometries are accessed directly, rather than *via* the excited state dynamics subsequent to a vertical excitation. As for the anisotropy parameter, the derived values for the VMP process are significantly larger than the values measured for one-photon photolysis.[107] This provides an additional, strong indication for the role of nonplanar geometries in the dynamics of the VMP of HN$_3$.[106]

References

1. B. J. Orr, *Int. Rev. Phys. Chem.*, 2006, **25**, 655.
2. J. Zhang, C. W. Riehn, M. Dulligan and C. Wittig, *J. Chem. Phys.*, 1995, **103**, 6815.
3. T. Arusi-Parpar, R. P. Schmid, R-J. Li, I. Bar and S. Rosenwaks, *Chem. Phys. Lett.*, 1997, **268**, 163.

4. R. P. Schmid, T. Arusi-Parpar, R-J. Li, I. Bar and S. Rosenwaks, *J. Chem. Phys.*, 1997, **107**, 385.

5. T. Arusi-Parpar, R. P. Schmid, Y. Ganot, I. Bar and S. Rosenwaks, *Chem. Phys. Lett.*, 1998, **287**, 347.

6. R. P. Schmid, Y. Ganot, I. Bar and S. Rosenwaks, *J. Chem. Phys.*, 1998, **109**, 8959.

7. R. P. Schmid, Y. Ganot, S. Rosenwaks and I. Bar, *J. Mol. Struct.*, 1999, **480–481**, 197.

8. I. Bar and S. Rosenwaks, *Int. Rev. Phys. Chem.*, 2001, **20**, 711.

9. Y. Ganot, X. Sheng, I. Bar and S. Rosenwaks, *Chem. Phys. Lett.*, 2002, **361**, 175.

10. X. Sheng, Y. Ganot, S. Rosenwaks and I. Bar, *J. Chem. Phys.*, 2002, **117**, 6511.

11. N. Yamakita, S. Iwamoto and S. Tsuchiya, *J. Phys. Chem. A*, 2003, **107**, 2597.

12. Y. Ganot, A. Golan, X. Sheng, S. Rosenwaks and I. Bar, *Phys. Chem. Chem. Phys.*, 2003, **5**, 5399.

13. A. Portnov, Y. Ganot, S. Rosenwaks and I. Bar, *J. Mol. Struct.*, 2005, **744–747**, 107.

14. Y. Ganot, S. Rosenwaks and I. Bar, *Z. Phys. Chem.*, 2005, **219**, 569.

15. T. Shimanouchi, *Tables of Molecular Vibrational Frequencies Consolidated Volume I*, National Bureau of Standards, Washington D.C., 1972.

16. D. J. Nesbitt and R. W. Field, *J. Phys. Chem.*, 1996, **110**, 12735.

17. M. Herman, J. Lievin, J. Vander Auwera and A. Campargue, *Adv. Chem. Phys.*, 1999, **108**, 1.

18. M. J. Bramley, S. Carter, N. C. Handy and I. M. Mills, *J. Mol. Spectrosc.*, 1993, **157**, 301.

19. A. Pisarchik, M. Abbouti Temsamani, J. Vander Auwera and M. Herman, *Chem. Phys. Lett.*, 1993, **206**, 343.

20. J. Lievin, M. Abbouti Temsamani, P. Gaspard and M. Herman, *Chem. Phys.*, 1995, **190**, 419.

21. C. K. Ingold and G. W. King, *J. Chem. Soc.*, 1953, **1953**, 2702.

22. K. K. Innes, *J. Chem. Phys.*, 1954, **22**, 863.

23. P. D. Foo and K. K. Innes, *Chem. Phys. Lett.*, 1973, **22**, 439.

24. C. Y. R. Wu, T. S. Chien, G. S. Liu, D. L. Judge and J. J. Caldwell, *J. Chem. Phys.*, 1989, **91**, 272.

25. A. M. Wodtke and Y. T. Lee, *J. Phys. Chem.*, 1985, **89**, 4744.

26. H. Shiromaru, Y. Achiba, K. Kimura and Y. T. Lee, *J. Phys. Chem.*, 1987, **91**, 17.

27. Y. Osamura, F. Mitsuhashi and S. Iwata, *Chem. Phys. Lett.*, 1989, **164**, 205.

28. B. A. Balko, J. Zhang and Y. T. Lee, *J. Chem. Phys.*, 1991, **94**, 7958.

29. D. H. Mourdaunt and M. N. R. Ashfold, *J. Chem. Phys.*, 1994, **101**, 2630.

30. Y.-C. Hsu, F.-T. Chen, L.-C. Chou and Y.-J. Shiu, *J. Chem. Phys.*, 1996, **105**, 9153.

31. M. N. R. Ashfold, D. H. Mourdaunt and S. H. S. Wilson, *Adv. in Photochem.*, 1996, **21**, 217.

32. Q. Cui, K. Morokuma and J. F. Stanton, *Chem. Phys. Lett.*, 1996, **263**, 46.

33. N. Hashimoto, N. Yonekura and T. Suzuki, *Chem. Phys. Lett.*, 1997, **264**, 545.

34. Q. Cui and K. Morokuma, *Chem. Phys. Lett.*, 1997, **272**, 319.

35. T. Suzuki, Y. Shi and H. Kohguchi, *J. Chem. Phys.*, 1997, **106**, 5292.

36. J-H. Wang, Y-T. Hsu and K. Liu, *J. Phys. Chem. A*, 1997, **101**, 6593.

37. Y. Shi and T. Suzuki, *J. Phys. Chem. A*, 1998, **102**, 7414.

38. D. H. Mourdaunt, M. N. R. Ashfold, R. N. Dixon, P. Loffler, L. Schnieder and K. H. Welge, *J. Chem. Phys.*, 1998, **108**, 519.

39. P. Loffler, E. Wrede, L. Schnieder, J. B. Halpern, W. M. Jackson and K. H. Welge, *J. Chem. Phys.*, 1998, **109**, 5231.

40. M. S. Child and L. Halonen, *Adv. Chem. Phys.*, 1984, **57**, 1.

40(a). T. P. Rakitzis, *Phys. Rev. Lett.*, 2005, **94**, 083005.

41. G. J. Scherer, K. K. Lehmann and W. Klemperer, *J. Chem. Phys.*, 1983, **78**, 2817.

42. A. Campargue, M. Abbouti Temsamani and M. Herman, *Mol. Phys.*, 1997, **90**, 793.

43. A. Campargue, M. Abbouti Temsamani and M. Herman, *Mol. Phys.*, 1997, **90**, 807.

44. A. Campargue, L. Biennier and M. Herman, *Mol. Phys.*, 1998, **93**, 457.

45. A. L. Utz, J. D. Tobiason, M. Carrasquillo, M. D. Fritz and F. F. Crim, *J. Chem. Phys.*, 1992, **97**, 389.

46. A. L. Utz, M. Carrasquillo, J. D. Tobiason, M. D. Fritz and F. F. Crim, *Chem. Phys.*, 1995, **190**, 11.

47. M. Abbouti Temsamani and M. Herman, *J. Chem. Phys.*, 1995, **102**, 6371.

48. M. Abbouti Temsamani and M. Herman, *J. Chem. Phys.*, 1996, **105**, 1355.

49. M. Herman, M. I. El Idrissi, A. Pisarchik, A. Campargue, A.-C. Gaillot, L. Biennier, G. D. Lonardo and L. Fusina, *J. Chem. Phys.*, 1998, **108**, 1377.

50. G. Herzberg, *Molecular Spectra and Molecular Structure III. Electronic Spectra and Electronic Structure of Polyatomic Molecules*, Van Nostrand Reinhold, New York, 1966.

51. J. Biesner, L. Schnieder, J. Schmeer, G. Ahlers, X. Xie, K. H. Welge, M. N. R. Ashfold and R. N. Dixon, *J. Chem. Phys.*, 1988, **88**, 3607.

52. J. Biesner, L. Schnieder, G. Ahlers, X. Xie, K. H. Welge, M. N. R. Ashfold and R. N. Dixon, *J. Chem. Phys.*, 1989, **91**, 2901.

53. D. H. Mourdaunt, M. N. R. Ashfold and R. N. Dixon, *J. Chem. Phys.*, 1996, **104**, 6460.

54. D. H. Mourdaunt, R. N. Dixon and M. N. R. Ashfold, *J. Chem. Phys.*, 1994, **104**, 6472.

55. R. V. Ambartzumian, V. S. Letokhov, G. N. Makarov and A. A. Puretzkiy, *Chem. Phys. Lett.*, 1972, **16**, 252.

56. A. Bach, J. M. Hutchison, R. J. Holiday and F. F. Crim, *J. Chem. Phys.*, 2002, **116**, 4955.

57. A. Bach, J. M. Hutchison, R. J. Holiday and F. F. Crim, *J. Chem. Phys.*, 2002, **116**, 9315.

58. H. Akagi, K. Yokoyama and A. Yokoyama, *J. Chem. Phys.*, 2003, **118**, 3600.

59. A. Bach, J. M. Hutchison, R. J. Holiday and F. F. Crim, *J. Chem. Phys.*, 2003, **118**, 7144.

60. A. Bach, J. M. Hutchison, R. J. Holiday and F. F. Crim, *J. Phys. Chem. A*, 2003, **107**, 10490.

61. M. L. Hause, Y. H. Yoon and F. F. Crim, *J. Chem. Phys.*, 2006, **125**, 174309.

62. H. Akagi, K. Yokoyama and A. Yokoyama, *J. Chem. Phys.*, 2004, **120**, 4696.

63. H. Akagi, K. Yokoyama, A. Yokoyama and A. Wada, *J. Mol. Spectrosc.*, 2005, **231**, 37.

64. G. Herzberg, *Molecular Spectra and Molecular Structure II. Infrared and Raman Spectra of Polyatomic Molecules*, Van Nostrand, Princeton, 1945.

65. D. R. Yarkony, *J. Chem. Phys.*, 2004, **121**, 628.

66. S. Koda and R. A. Back, *Can. J. Chem.*, 1977, **55**, 1380.

67. A. Nakajima, K. Fuke, K. Tsukamoto, Y. Yoshida and K. Kaya, *J. Phys. Chem.*, 1991, **95**, 571.

68. S. S. Brown, H. L. Berghout and F. F. Crim, *J. Chem. Phys.*, 1995, **102**, 8440.

69. F. F. Crim, *J. Phys. Chem.*, 1996, **100**, 12725.

70. S. S. Brown, R. B. Metz, H. L. Berghout and F. F. Crim, *J. Chem. Phys.*, 1996, **105**, 6293.

71. S. S. Brown, H. L. Berghout and F. F. Crim, *J. Chem. Phys.*, 1996, **105**, 8103.

72. S. S. Brown, C. M. Cheatum, D. A. Fitzwater and F. F. Crim, *J. Chem. Phys.*, 1996, **105**, 10911.

73. S. S. Brown, H. L. Berghout and F. F. Crim, *J. Chem. Phys.*, 1997, **106**, 5805.

74. J.-J. Klossika, H. Flöthmann, C. Beck, R. Schinke and K. Yamashita, *Chem. Phys. Lett.*, 1997, **276**, 325.

75. S. S. Brown, H. L. Berghout and F. F. Crim, *J. Chem. Phys.*, 1997, **107**, 8985.

76. S. S. Brown, H. L. Berghout and F. F. Crim, *J. Chem. Phys.*, 1997, **107**, 9764.

77. H. L. Berghout, S. S. Brown, R. Delgado and F. F. Crim, *J. Chem. Phys.*, 1998, **109**, 2257.

78. M. J. Coffey, H. L. Berghout, E. Woods and F. F. Crim, *J. Chem. Phys.*, 1999, **110**, 10850.

79. J.-J. Klossika and R. Schinke, *J. Chem. Phys.*, 1999, **111**, 5882.

80. H. L. Berghout, F. F. Crim, M. Zyrianov and H. Reisler, *J. Chem. Phys.*, 2000, **112**, 6678.

81. E. Woods, H. L. Berghout, C. M. Cheatum and F. F. Crim, *J. Phys. Chem. A*, 2000, **104**, 10356.

82. R. Schinke and M. Bittererová, *Chem. Phys. Lett.*, 2000, **332**, 611.

83. H. L. Berghout, S. Hsieh and F. F. Crim, *J. Chem. Phys.*, 2001, **114**, 10835.

84. K. Yamada, *J. Mol. Spectrosc.*, 1980, **79**, 323.

85. G. Herzberg and C. Reid, *Discuss. Faraday Soc.*, 1950, **9**, 92.

86. A. L. L. East, C. S. Johnson and W. D. Allen, *J. Chem. Phys.*, 1993, **98**, 1299.

87. S. S. Brown, H. L. Berghout and F. F. Crim, *J. Phys. Chem.*, 1996, **100**, 7948.

88. R. A. Brownsword, T. Laurent, R. K. Vatsa, H. R. Volpp and J. Wolfrum, *Chem. Phys. Lett.*, 1996, **258**, 164.

89. R. A. Brownsword, M. Hillenkamp, H. R. Volpp and R. K. Vatsa, *Res. Chem. Intermed.*, 1999, **25**, 339.

90. W. K. Yi and R. Bersohn, *Chem. Phys. Lett.*, 1993, **206**, 365.

91. L. J. Butler, T. M. Ticich, M. D. Likar and F. F. Crim, *J. Chem. Phys.*, 1986, **85**, 2331.

92. T. M. Ticich, M. D. Likar, H-R. Dübal, L. J. Butler and F. F. Crim, *J. Chem. Phys.*, 1987, **87**, 5820.

93. M. Brouard, M. T. Martinez, J. O'Mahony and J. P. Simons, *Chem. Phys. Lett.*, 1988, **150**, 6.

94. M. D. Likar, J. E. Baggott, A. Sinha, T. M. Ticich, R. L. Vander Wal and F. F. Crim, *J. Chem. Soc., Faraday Trans. 2*, 1988, **84**, 1483.

95. M. D. Likar, A. Sinha, T. M. Ticich, R. L. Vander Wal and F. F. Crim, *Ber. Bunsenges. Phy. Chem.*, 1988, **92**, 289.

96. M. Brouard, M. T. Martinez, J. O'Mahony and J. P. Simons, *J. Chem. Soc., Faraday Trans. 2*, 1989, **85**, 1207.

97. M. Brouard, M. T. Martinez, J. O'Mahony and J. P. Simons, *Mol. Phys.*, 1990, **69**, 65.

98. M. Brouard, M. T. Martinez and J. O'Mahony, *Mol. Phys.*, 1990, **71**, 1021.

99. M. Brouard, M. T. Martinez, J. O'Mahony and J. P. Simons, *Philos. Trans. R. Soc. Lond. A*, 1990, **332**, 245.

100. D. T. Colbert and E. L. Sibert III, *J. Chem. Phys.*, 1991, **94**, 6519.

101. M. Brouard and R. Mabbs, *Chem. Phys. Lett.*, 1993, **204**, 543.

102. F. F. Crim, *Annu. Rev. Phys. Chem.*, 1993, **44**, 397.

103. Y. Guan and D. L. Thompson, *Chem. Phys.*, 1989, **139**, 147.

104. S. M. Holland, R. J. Sticklandt, M. N. R. Ashfold, D. A. Newnham and I. M. Mills, *J. Chem. Soc., Faraday Trans.*, 1991, **87**, 3461.

105. R. J. Barnes, A. F. Gross, M. Lock and A. Sinha, *J. Phys. Chem. A*, 6133, **101**, 1997.
106. R. J. Barnes, A. Sinha, P. J. Dagdigian and H. M. Lambert, *J. Chem. Phys.*, 1999, **111**, 151.
107. U. Meiert and V. Staemmler, *J. Phys. Chem.*, 1991, **95**, 6111.
108. K-H. Gericke, M. Lock, R. Fasold and F. J. Comes, *J. Chem. Phys.*, 1992, **96**, 422.

VMP of Larger than Tetratomic Molecules

As stated at the beginning of Chapter 7, VMP theories and experiments that deal with tetratomic molecules are much more complex than for triatomics. The arguments made there certainly hold when we deal with VMP of even larger molecules and we may anticipate a much higher complexity for this "class" of species. Moreover, as explained below (Section 8.1), it is anticipated that IVR in these molecules would preclude mode-selective or bond-selective dissociation in the ns timescale. Apparently, these difficulties were not fully appreciated in the early days of VMP and in 1972 it was suggested to apply VMP for dissociating selectively diethermethylketone in a mixture that also contains dietherketone.[1] It seems that an experimental verification of this scheme was not tried (or failed). Indeed, at present only for one "larger than tetratomic" molecule, methylamine, was mode-dependent enhancement in photodissociation observed.

Before discussing the VMP of the title molecules, where specific vibrational levels were pre-excited, we note that low-frequency vibrational modes, typical of large molecules, are significantly populated even at fairly low temperatures and may affect the course of their photodissociation. Although we will not discuss this scenario here, we cite one example, the recent work on photodissociation of thiophenol, which suggests a temperature dependence of the photofragments branching ratio.[1a]

We start this chapter with presentation of the VMP of methylamine isotopologues (Section 8.1) and then move on to haloalkanes where, in some instances, vibrational pre-excitation changes the electronically excited states accessed in photodissociation and the branching ratio between the atomic photofragments (Section 8.2). The subsequent section (8.3) deals with phenol, where the dynamics of CIs in this large molecule is exemplified *via* comparison of the recoil-energy distributions of the fragments from one-photon

Vibrationally Mediated Photodissociation
By Salman (Zamik) Rosenwaks
© Salman (Zamik) Rosenwaks 2009
Published by the Royal Society of Chemistry, www.rsc.org

dissociation with those from VMP. The last section (8.4) reports on VMP of other "larger than tetratomic" molecules.

8.1   VMP of Methylamine Isotopologues

8.1.1   Spectral and Dynamical Features Relevant to VMP of Methylamine

The interest in VMP of methylamine, CH_3NH_2, the simplest primary amine, and its deuterated isotopologues, stems from the fact that its excitation *via* the $\tilde{A} \leftarrow \tilde{X}$ transition (the first absorption band at ~ 200–245 nm[1b]) may lead to N–H(D), C–H(D) or C–N bond rupture. The thermodynamically accessible dissociation channels following excitation in this band are depicted in Figure 8.1.

In one-photon dissociation, N–H(D) rupture was found to be the major channel.[2–10a] It is of interest to examine the possible effect of specific vibrational pre-excitation on the efficacy of photodissociation and on the branching ratio between the photofragments. However, this is not an easy task since methylamine is a seven-atom molecule with fifteen vibrational NMs. The frequencies of these modes range from 268 cm^{-1} for ν_{15}, the CH_3 torsion mode, to 2820–3427 cm^{-1} for the various CH_3 and NH_2 stretches in CH_3NH_2, with lower

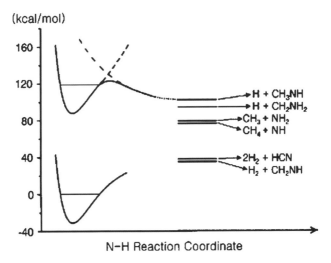

Figure 8.1   Simple diagram of PESs along the N–H(D) bond dissociation coordinate. Thermodynamically accessible reaction channels are shown on the right. For the upper state, the adiabatic PES is depicted in the vicinity of the FC region. Thereafter, it correlates to the $H + CH_3NH(\tilde{X})$ products diabatically in the late stage along the reaction coordinate. Reprinted with permission from Ref. 9. Copyright 2003, American Institute of Physics.

frequencies in the isotopologues for modes where D atom movements are involved.[11] In particular, it exhibits two strongly coupled large-amplitude motions, comprising the torsion of the methyl top and the inversion of the amine group[12,13] with barriers for torsion and inversion of ~ 690,[10] and $1943\,cm^{-1}$,[13] respectively. Thus, mode specificity in VMP is expected to be more difficult to observe than in tri- or tetratomic molecules due to fast energy flow out of the initially prepared states. As explained below, this is particularly true for molecules with a torsional degree of freedom, which are expected to exhibit fast IVR.[14] Nevertheless, it turns out that notwithstanding these "expectations", mode-dependent enhancement was observed in the VMP of methylamine and a large part of the rest of this section is devoted to this observation.

Before going into the details of methylamine VMP investigations[15–20] it is useful to briefly review studies on its one-photon dissociation. In the photolysis of methylamine with broadband radiation (194–244 nm),[2] four major dissociation pathways were observed: the dominant channel corresponding to N–H bond fission accounting for at least 75% of the dissociation, C–H bond rupture for ~ 7.5%, molecular-hydrogen elimination from the amine for less than 10% and very little CN bond breaking yielding methyl and amine radicals. More recently, investigation of the photodissociation of methylamine under collisionless conditions at 222 nm in a crossed laser-molecular beam[4] led to the conclusion that the major channel is N–H bond fission, while the C–N and C–H bond dissociation channels and the H_2 elimination are also significant. In addition, photodissociating four isotopologues of methylamine, CH_3NH_2, CH_3ND_2, CD_3NH_2 and CD_3ND_2 in the 203–236 nm wavelength region, followed by H(D) HRTOF PTS,[6,7] led to the conclusion that most if not all slow and fast H(D) photofragments arise from N–H(N–D) bond fission. Both contributions were considered to arise from H-atom tunneling through (or passing over) an early barrier in the NH dissociation coordinate and evolution into the region of the CI between the $\tilde{A}$ and $\tilde{X}$ state surfaces. The fast hydrogen atoms were linked to molecules that pass directly through the CI to the $H + CH_3NH(\tilde{X})$ asymptote, while the slow atoms were attributed to those molecules that "miss" the CI on the first traversal but make the $\tilde{X} \leftarrow \tilde{A}$ transfer on a later encounter.

In more recent studies of one-photon dissociation of methylamine,[8–10a] severe homogeneous broadening was observed in $CH_3NH_2(\tilde{A}, v')$ states corresponding to lifetimes of a fraction of a ps. Replacement of the NH_2 moiety by ND_2 resulted in a large increase of the lifetime, whereas replacement of CH_3 by CD_3 did not result in a similar effect. These observations strongly indicate that the product H atom is obtained primarily *via* N–H bond breaking.[8–10a] Moreover, partial H to D substitution on the amino moiety was not effective in reducing the linewidth of the vibronic band, indicating that the predissociation occurs exclusively along the N–H coordinate with a reaction barrier through which the tunneling rate leading to the N–D bond rupture is significantly slowed down due to the larger mass of D.[10,10a]

1 Evidence for Mode-Dependent Enhancement of Bond Fission and Photoionization of CH_3NH_2

8.2 displays the room-temperature PARS spectrum, the jet-cooled spectrum, the REMPI spectrum of CH_3NH_2, and, on the right side of anel, the respective excitation schemes.[5] The spectra are characterized by ple-peak structure, related to the Q-branches of different bands, and the of the action show up whenever the difference frequency of the SRE laser matches that of a specific vibrational transition. However, whereas in ARS spectrum two of the peaks, of the degenerate CH_3 stretch, v_2 m^{-1}) and the CH_3 symmetric stretch, v_3 (2820 cm^{-1}), are dominant and hers quite weak, in the action spectrum all peaks carry significant ty and, in particular, the PARS low-intensity peaks become prominent. litional dominant peak, in both the PARS and action spectra, due to the ymmetric stretch, v_1 (3361 cm^{-1}), is beyond the wave-number span of the

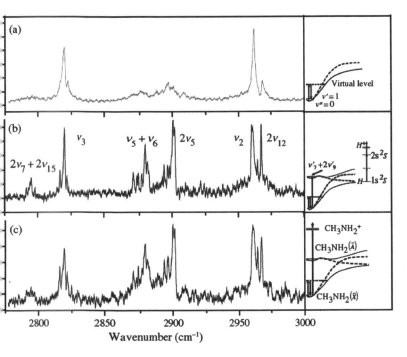

8.2 Vibrational excitation of the CH_3 stretches region of CH_3NH_2: (a) the room-temperature PARS, (b) the jet-cooled action spectrum monitoring the REMPI yield of H photofragments following 243 135-nm photo-dissociation of pre-excited CH_3NH_2 and (c) the REMPI spectrum of CH_3NH_2. The insets at the right display the excitation scheme used to obtain each spectrum; the PESs for N–H dissociation are adapted from Ref. 5, the dashed line denotes C_s symmetry and the solid line C_1 symmetry. Reproduced from Ref. 17 with permission.

These experimental findings are in line with theore
ab initio calculations[3] showed small barriers in the excited
C–N and N–H fission reaction coordinates. It was argue
C–N bond is weaker than the N–H bond, in the excited sta
fission on the $\tilde{A}$ state surface is higher than that for N–H
more efficient bond breaking of the latter upon excitatio
state. A later analysis[5] put forward the explanation that ex
lone-pair electron into a Rydberg *s* orbital produces an ex
with other electronic states and decomposes *via* N–H, C
Cuts through *ab initio* PESs for the $\tilde{X}$ and $\tilde{A}$ states of CH
indicating that the $\tilde{A}$ state potential for both N–H and
characterized by a small barrier at short range. For N–H
excited-state and ground-state surfaces cross as the
forming a CI. The N–H rupture excited-state products li
energy and dissociating molecules are funneled through
ground-state products. The C–N rupture excited-state p
excitation energy and dissociation would proceed mostl
surface. In the case of C–H fission, the excited-state p
excitation energy, but since there is no CI in this channe
occur only after IC to a hot ground state, with a large fra
energy partitioned to internal motions of the product:
translational kinetic energy.[5]

Although the above conclusions on one-photon dis
guide for interpreting VMP processes in methylamine,[15]
tion dynamics in the latter case may differ and, as is obvi
examples, further experimental and theoretical studies
understand these complex processes. We note the latest
the spectral and dynamical features of methylamine tha
density functional theory (DFT) calculations for o
vibrational frequencies in CH_3NH_2, CH_3ND_2, CD_3NH_2
detailed calculations of CIs in CH_3NH_2.[20b]

8.1.2 VMP of CH_3NH_2 in the Region of the
and NH_2 Stretches

Excitation of CH_3[15–17] and NH_2[16] fundamental stretc
isoenergetic combinations and overtones of CH_3 deforr
the 2770–3400 cm^{-1} range was achieved *via* SRE. The e
~243.1-nm excitation on the H atom dissociation chan
monitoring the jet-cooled action spectrum, which reflec
the H photofragments as a function of vibrational excita
to the PARS spectrum. In addition, H-atom Doppler p
by parking the SRE wavelengths on the *Q*-branch i
transitions to specific vibrational states) and tuning the
REMPI transition.

8.

Fig
act
eac
a n
pea
bea
the
(29
the
inte
An
NH
figu

Intensity (arb. units)

Figu

The bands at 2795, 2880, 2901 and 2968 cm^{-1} were tentatively assigned to $2v_7 + 2v_{15}$, $v_5 + v_6$, $2v_5$ and $2v_{12}$, respectively,[15] where v_5 (1473 cm^{-1}) corresponds to the CH$_3$ degenerate deformation, v_6 (1430 cm^{-1}) to the CH$_3$ symmetric deformation, v_7 (1130 cm^{-1}) to the CH$_3$ rock, v_{12} (1485 cm^{-1}) to the CH$_3$ degenerate deformation and v_{15} (268 cm^{-1}) to the CH$_3$ torsion. Thus, the features related to the deformations are of higher intensity, meaning that they exhibit higher enhancement in the action spectrum relative to the CH$_3$ and NH$_2$ stretches. An estimate of the enhancement was obtained from the ratio between the peak areas corresponding to the same feature in the action and PARS spectra, setting that of v_2 to 1.0 and normalizing the other features' intensity to it. The enhancement factors for the NH$_2$ and CH$_3$ stretches (v_1, v_2 and v_3) are 0.9–1.2, while those corresponding to the CH$_3$ deformation bands are 1.7–2.5. This behavior suggests higher enhancement for the deformations and thus mode-dependent enhancement and promotion of H photofragmentation in the $\sim$243.1-nm VMP of CH$_3$NH$_2$.[15–17]

The H-photofragment yield depends on the SRE vibrational excitation probability, the FC overlap between the vibrational wavefunctions in the ground and excited electronic states, the electronic transition dipole moment and on the photodissociation channel. Since in the VMP process the combined SRE + UV excitation energies are in the $\sim$43 900–44 530 cm^{-1} range, the CH$_3$NH$_2$ $\tilde{A}$ state is accessed from all the initially prepared vibrational states. It was therefore concluded[15–17] that the main player determining the H-yield dependence is the FC factor in the UV excitation from the different pre-excited vibrational states. The differences in the FC factors can be rationalized by noting that the $\sim$243.1-nm excitation of the CH$_3$ stretches or overtones and combinations of deformations at 43 900–44 100 cm^{-1} access the $v'_7 2v'_9$ homogenously broadened level ($\sim$200 cm^{-1}) on the $\tilde{A}$ PES (v'_7 is the CH$_3$ rock and v'_9 the NH$_2$ wag).[9] This suggests that the initially prepared deformations overlap better with the upper state and result in FC factors larger than for the CH$_3$ stretch states.

This conclusion was corroborated by monitoring the CH$_3$NH$_2$ REMPI spectrum from the same initial vibrational states, using temporally overlapping SRE and UV photons.[15–17] Figure 8.2(c), shows that the molecular REMPI spectrum, which directly probes the electronic transition, has a similar pattern to that of the H action, Figure 8.2(b). The resemblance of the two types of spectra suggests that both are driven by the same FC factors.

As reported in the previous chapters, similar behavior of the action spectra was encountered in smaller molecules, *e.g.*, C$_2$H$_2$, C$_2$HD and HNCO, where excitation of combination bands including bending, affected considerably the FC factors and subsequently the bond fission. Nevertheless, mode specificity should be less likely in larger molecules, due to faster energy flow out of the initially prepared states. This is particularly true for molecules with a torsional degree of freedom. It was pointed out that IVR is accelerated in flexible molecules when the prepared vibration is close to the center of flexibility.[14] In particular, the dependence of IVR rates on the internal rotation barrier height was carefully studied in four- to twelve-atom molecules.[14] It was shown that there is a systematic decrease in IVR lifetimes as the barrier is decreased from $1700\,\mathrm{cm}^{-1}$ to $400\,\mathrm{cm}^{-1}$. Following the conclusions of that study it is inferred that CH_3NH_2, which is characterized by a torsion barrier of $\sim 690\,\mathrm{cm}^{-1}$,[10] ought to have IVR lifetimes < 200 ps. However, the VMP results presented here clearly show that the initially prepared states live considerably longer, sustaining their nuclear motion and leading to favorable FC factors and thus to enhanced dissociation. *Were IVR complete on the ns timescale of the present experiment (where there is ~ 10 ns delay between the SRE and UV photons), one might expect FC factors from either initially prepared states to be identical. In contrast, the differing FC factors point to a persisting difference in the character of the vibrationally excited states.*

8.1.2.2 H-Atom Doppler Profiles: Evidence on the Main Photodissociation Channel

As explained at the beginning of Section 8.1, N–H(D) rupture was concluded to be the major channel in one-photon dissociation of methylamine *via* the first absorption band. Similar conclusions on the dominant photodissociation channel were obtained from the Doppler profiles of the H atoms produced following the ~ 243.1-nm excitation of CH_3NH_2 pre-excited to the NH_2 symmetric stretch fundamental (v_1) and the CH_3 symmetric stretch fundamental (v_3).[16] The Doppler profiles for these excitations are depicted in Figure 8.3. The profiles resulting from other, approximately isoenergetic pre-excited vibrational states were very similar with widths between those for v_1 and v_3. The measured profiles (squares) are well fitted by Gaussians (solid lines in Figures 8.3(a) and (b)), suggesting isotropic velocity distribution. The Gaussian profiles correspond to Maxwellian velocity distributions and allow finding the translational-energy distributions of the H photofragments (Figures 8.3(c) and (d)). The widths of the H Doppler profiles in photodissociation of pre-excited NH_2 and CH_3 stretches correspond to average translational energies of the H photofragment of 0.42 ± 0.02 and 0.30 ± 0.03 eV, respectively. These energies are considerably lower then the maximum available energies of about 1.2 and 1.7 eV (marked by arrows in Figure 8.3) for release of H atoms from the NH_2 and CH_3 moieties (obtained by subtracting the dissociation energies of 4.28 and 3.79 eV for the N–H and C–H pathways from the combined SRE + UV energies of 5.51 and 5.45 eV, respectively). The fact that the yield of H atoms (inferred from the action spectra) for the two excitations is about the same and that the difference between the average translational energies is only ~ 0.1 eV, whereas between the available energies it is ~ 0.5 eV, suggests that (mostly) the same bond is ruptured in both cases.

The available energies for N–H and C–H bond cleavage marked by the two arrows in Figures 2(c) and (d) can also be compared to the distribution of the translational energies of the H photofragments. It is clearly seen that the arrows for the N–H and C–H cleavage are positioned on the tails of the energy-distribution curves. Since there are barely any H atoms carrying energy higher than 1.2 eV (the available energy for N–H bond cleavage), it is unreasonable that the C–H channel is the dominant one.

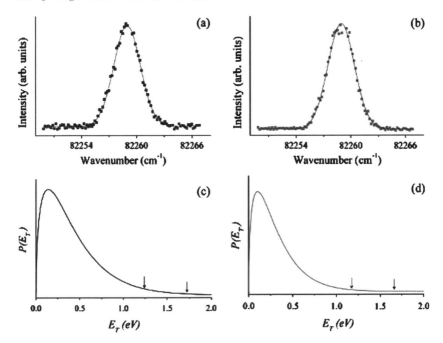

Figure 8.3 Representative H Doppler-broadened profiles obtained *via* ∼243.1-nm photodissociation of methylamine initially excited to the fundamental (a) NH_2 symmetric stretch, (v_1), and (b) CH_3 symmetric stretch, (v_3). The solid curves are the Gaussians fitted to the corresponding measured profiles (squares). (c) and (d) are the Maxwellian translational-energy distributions extracted from (a) and (b) with the arrows marking the maximum available translational energies of H as a result of N–H and C–H bond fission, respectively. Reproduced from Ref. 16 with permission. Copyright 2007 LPP Ltd.

The fraction of energy released into H translation, calculated assuming that the dominant channel is N–H, is 0.35 and 0.25 for ∼243.1-nm dissociation of CH_3NH_2 initially excited to the NH_2 and CH_3 stretches, respectively. This means that relatively small fractions of translational energies are channeled into the H photofragments, pointing to a nonadiabatic dissociation process. This is in line with the explanation given above, suggesting that the H atoms tunnel through a barrier in the N–H dissociation coordinate and funnel through the CI between the $\tilde{A}$ and $\tilde{X}$ state surfaces to the (hot) ground electronic state where they dissociate.

8.1.3 VMP of CH_3NH_2 in the Region of the NH_2 Overtone Stretches

Nonstatistical energy flow in the first NH_2 symmetric stretch overtone $(2v_1)$ region was observed in CH_3NH_2.[18,20] This was concluded from measurements of jet-cooled action spectra[18,20] and Doppler profiles,[20] and of room-temperature, PA vibrational excitation spectra.[18,20] Action spectra of ~ 243.1-nm excitation of the pre-excited first to third NH_2 stretching overtones[20] showed partially resolved features, particularly in the first NH_2-stretch overtone region. The simulation of these spectra together with the PA spectra allowed retrieval of the band origins, types and intensities, and the homogeneous linewidths. By modeling these data with a simplified LM/NM and NM models and taking into account the Fermi resonances, it was possible to assign the observed features and show that they are related to couplings of the bright and doorway states in the amino group.[20] Furthermore, by considering the LM/NM Hamiltonian and the retrieved transition linewidths it was possible to decode the intramolecular dynamical information. Of particular note is that the antisymmetric NH_2 stretch state ($2v_{10}$) is relatively long lived, ~ 4.4 ps, up to about three times longer than other, adjacent states, implying nonstatistical energy flow. This might be due to the average coupling of the $2v_{10}$ state to the bath states being weaker, or the accidental absence of low-order resonances, leading to a bottleneck in the initial stages of relaxation and consequently to a slower decay process.

The Doppler profiles and the extracted translational-energy distributions indicated that the translational energy disposal in the H photofragments is low, and even more so, it is lower for ~ 243.1-nm dissociation of CH_3NH_2 excited with three and four quanta of NH_2 than with two quanta. This behavior was attributed to a change in the dissociation mechanism. The photofragmentation in the latter was suggested to occur *via* IC to the (hot) ground-state surface, while in the former *via* both the parent $\tilde{A}$ and $\tilde{X}$ states, where the $\tilde{A}$ state adiabatically correlates with an H atom and an electronically excited $CH_3NH(\tilde{A})$.[20]

8.1.4 VMP of CD_3NH_2

VMP of partially deuterated methylamine, where the methyl and amine moieties consist of different isotopes of hydrogen, could help in discriminating between C–H(D) and N–H(D) bond rupture and for assessing possible mode-selective bond fission. To this end, jet-cooled H and D action spectra and Doppler profiles, obtained through dissociation at ~ 243.1 nm of CD_3NH_2 initially excited to the first NH_2 symmetric stretch overtone region, were studied.[19] The monitored action spectra were somewhat narrower than the room-temperature PA spectrum, due to the inhomogeneous structure reduction, exhibiting some residual structure. The simulation of the action spectra allowed identification of the contribution of multiple bands to the dominant feature and retrieving of the band types, origins and homogeneous linewidths. The assignment of these bands enabled determination of the vibrational pattern and involved resonances, and decoding of the intramolecular dynamical characteristics. It was shown that Fermi resonances are of importance in this range, but also higher-order resonances contribute to the vibrational pattern of CD_3NH_2. The dynamical picture was retrieved by finding the positions of the zero-order states and the homogeneous linewidths, enabling the timescales for energy flow to the interacting doorway and bath states to be obtained. As in CH_3NH_2 (see previous section), the relatively isolated $2v_{10}$ state was found to have a lifetime of ~ 4.4 ps, up to about three times longer than those of other states in this energetic window, indicating nonstatistical energy redistribution or some state specificity, even at the high density of states in the vicinity.

From the areas of the H and D Doppler profiles, the H/D branching ratio is 9/1 and from the extracted translational-energy distribution, the average translational energy of the H photofragments is higher, 0.47 ± 0.02 eV, than that of D, 0.38 ± 0.01 eV. Although the specificity attained by excitation of the NH_2 stretch overtones was lost in picoseconds, H atoms were still produced preferentially. This was attributed to different mechanisms of CD_3NH_2 dissociation in the $\tilde{A}$ state. The topology of the PESs, the transition states on the N–H dissociation coordinate being lower than on the C–D (as calculated for C–H[5]) and the combined IR + UV excitation energy, explains the dominance of the N–H rupture. The extracted translational-energy distributions of the H and D products indicate that the former are associated with dissociation involving motion on the N–H coordinate on the excited $\tilde{A}$ surface, tunneling through an early barrier and evolvement into the region of the CI to the ground-state surface, followed by dissociation to ground-state H and CD_3NH products. The lower translational energy of the D product agrees with dissociation *via* IC to the ground state and decay to D and vibrationally excited CD_2NH_2 products.[5]

8.2   VMP of Haloalkanes

In the photodissociation of haloalkanes in the first absorption band, halogen atoms are mostly the main atomic photofragments, but H atoms (D in isotopologues) are detected as well in many cases. Haloalkanes are denoted here as $C_lH_mX_n$, where $l = 1,2 \ldots$, $m = 0,1,2 \ldots$ and X_n represents $n = 1,2 \ldots$ halogen atoms (it may represent different numbers of different halogen atoms in the same molecule). The ensuing X atoms are observed in the two spin-orbit states of the ground electronic state: the lower, X $[\equiv X(^2P_{3/2})]$ and higher, X* $[\equiv X(^2P_{1/2})]$ state. For reasons explained below, many VMP studies of haloalkanes were carried out on hydrochlorofluorocarbon (HCFC) molecules. Therefore, and since some specific effects of vibrational preparation on the photodissociation dynamics of these molecules were observed, a large part of the present section is dedicated to HCFCs (Section 8.2.1). VMP of methyl iodide, CH_3I, is also dealt with in some detail, due to its benchmark role in studying multichannel dissociation of simple polyatomic molecules (Section 8.2.2). Other haloalkanes are reviewed in less detail (Section 8.2.3).

8.2.1   Hydrochlorofluorocarbons

The introduction of HCFCs as interim replacements to the ozone-destroying chlorofluorocarbon (CFC) and halon molecules[21,22] produced considerable interest in their photochemistry.[23-30] The HCFCs, containing at least one C–H bond, were pursued as substitutes since they have shorter atmospheric lifetimes than the CFCs and halons due to the reaction of the C–H moiety with OH and $O(^1D)$. Studies of one-photon dissociation of HCFCs established that they release mainly Cl atoms and to some extent H atoms during their photo-fragmentation in the first absorption band (180–240 nm).[24-29]

In the VMP studies of HCFCs presented below, $CHFCl_2$, CHF_2Cl, CH_3CFCl_2 and CH_3CF_2Cl were interrogated.[31-38] Vibrational excitation in the regions of the fundamental (using SRE) and several CH_n overtone stretches was followed by UV excitation. Room-temperature PA and jet-cooled action spectra of Cl, Cl* and H photofragments were monitored. (2 + 1) REMPI detection of the photofragments was employed at ~ 235 nm for the two Cl species and at ~ 243 nm for H. The UV excitation wavelength was the same as that used for REMPI in the specific experiment. The spectra represent the contributions of the different isotopologues of the above species according to the abundance of the ^{35}Cl and ^{37}Cl isotopes.

Each of the above four species has specific VMP features, However, those of the two methane derivatives, $CHFCl_2$[33,34,36] and CHF_2Cl,[37] and of the ethane derivatives, CH_3CFCl_2[31,32,35,36] and CH_3CF_2Cl,[38] have somewhat similar spectroscopy (linewidths, lifetimes, enhancement of photofragment production). We therefore dwell on the species where most of the work was done, $CHFCl_2$ and (mainly) CH_3CFCl_2.

8.2.1.1   *CHFCl₂*

(a) Enhanced Production of Photofragments.   PA and action spectra for $N=3$, $N=7/2$, and $N=4$ CH stretch-bend polyads of $CHFCl_2$ were observed.[33] The spectra are assigned and labeled in the polyad notation applying a model that involves coupling between the CH stretching and the CH bending motions.[39,40] Each polyad is assigned by its quantum number N with its components labeled as $N_j = v_s + 1/2\ v_a + 1/2\ v_b$, where v_s counts the number of zero-order bright stretches, v_a and v_b of the first and second bending, and the subscript j labels the components of the polyad starting from the highest frequency (*i.e.*, the components are combinations of different vs that sum up to N). For $CHFCl_2$,[40] the CH stretch ($v_1 = 3024.8$ cm^{-1}) is anharmonically coupled by the 1:2 Fermi resonance with either of the two CH bends, $v_2 = 1317.2$ cm^{-1} or $v_7 = 1242.6$ cm^{-1} and the two bends are coupled by a Darling–Dennison resonance. These resonances lead to a mixing of states that form the polyad components N_j.

The action spectra exhibited significant increase in Cl, Cl* and H photo-products yield as a result of VMP, relative to the background one-photon photodissociation.[33] By measuring the yields of the different photofragments and particularly the branching ratios between them, some features of the photodissociation dynamics could be revealed. Comparison of the total Cl yield, in both spin-orbit states, to that of H showed that the former was enhanced four times more than the latter for the VMP of the 3_1, $7/2_2$ and 4_1

components.[33] This finding emphasizes that although the C–H stretch was initially extended, the cleavage of the C–Cl bond was more efficient. This agrees with analysis that showed that the vibrational energy redistribution in $CHFCl_2$ for the $N = 7/2$ polyad occurs on a timescale of ps.[33] Thus, for the ns laser pulse widths employed in this experiment the mixed states were prepared and the energy was already redistributed into the modes corresponding to the heavy atoms during the excitation step. Extension of the C–Cl bond, which becomes the reaction coordinate, may lead to a favorable FC factor with the upper PESs, probably due to the absorption characteristics related to a $\sigma^*_{C-Cl} \leftarrow n$ transition.[40,41]

By accounting for the transition probability ratios of the tagged fragments[41–44] the REMPI signals were converted into Cl*/Cl and H/(Cl + Cl*) branching ratios. The branching ratios for Cl*/Cl in VMP of $CHFCl_2$ was close to 0.5 independent of the prepared polyad component. The H/(Cl + Cl*) was in the range of 0.11–0.15, pointing to low fraction of produced H relative to that of Cl and Cl*.[33]

(b) Evidence for Vibrationally Induced Three-Body Photodissociation. The ^{35}Cl and $^{35}Cl^*$ ion arrival-time profiles following the $\sim$235-nm photodissociation of $CHFCl_2$ prepared in the Q-branch (comprising several unresolved rotational states) of the 3_1, 4_1 and 5_1 stretch-bend polyads were measured[34] and are displayed in Figure 8.4. These profiles represent the VMP "net" profiles, obtained by removing the contribution, when significant, resulting from the $\sim$235-nm photodissociation of vibrationless ground-state molecules (vibrational excitation laser off) from the signal monitored when both the vibrational excitation laser and the UV laser were on.[34] The observed profiles were obtained with the polarization of the photolysis/ probe UV laser parallel or perpendicular to the TOFMS axis and with the polarization of the IR/VIS vibrational excitation laser axis remaining fixed with perpendicular polarization. The profiles of both Cl and Cl*, taken under these polarization conditions, are shown in Figures 8.4(a)–(c). Changing the polarization of the IR/VIS laser (including from perpendicular to parallel to the TOFMS axis) did not affect the shape of profiles. This indicates that the vibrational excitation to the Q-branch of the 3_1, 4_1 and 5_1 states does not induce significant alignment to the molecules.

The Cl and Cl* photofragment spectra, shown in Figures 8.4(a) and (b), indicate that the profiles are doubly peaked for the parallel polarization of the UV laser and singly peaked for the perpendicular one. The double peaks are due to formation of photofragments of equal translational energies but with velocity vectors pointing toward and opposite to the flight axis. The profiles shapes indicates that both Cl and Cl* photofragments are released predominantly through a parallel electronic transition with a positive β.[45–47] Also, from comparison of the profiles of Cl and Cl* obtained in VMP *via* the intermediate state 5_1 to those *via* 3_1 and 4_1 it is evident that the intensities of the centers of the arrival-time profiles increase and are much more pronounced for 5_1. This happens although the profiles are monitored under similar conditions,

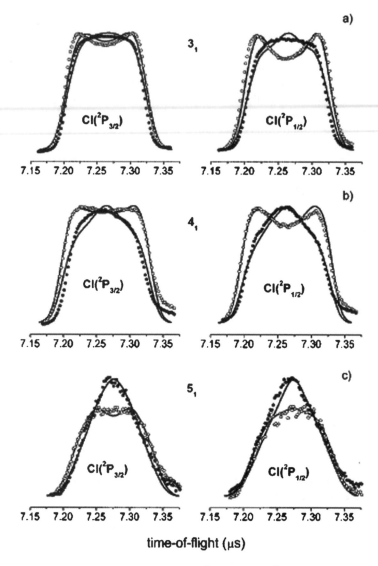

Figure 8.4   Arrival-time distributions of $Cl(^2P_{3/2})$ and $Cl(^2P_{1/2})$ photofragments produced in the $\sim$ 235-nm photolysis of $CHFCl_2$ pre-excited to the Q-branch
of the a) 3_1, b) 4_1 and c) 5_1 polyads. Open circles and solid points are the
experimental data points taken with the polarization of the UV photolysis/
probe laser parallel and perpendicular, respectively, to the TOFMS axis.
The polarization of the overtone excitation laser was perpendicular to the
TOFMS axis. Solid lines are the simulations of the corresponding profiles.
These lines denote the best-fit velocity distributions, with constant β, finite
time response of the apparatus (modeled as a Gaussian with 20 ns full
width half-maximum) and Doppler selection by the finite bandwidth of
the probe laser (modeled as $0.3\,\mathrm{cm}^{-1}$ at the one-photon wave number).
Reprinted with permission from Ref. 34. Copyright 2001, American
Institute of Physics.

except the excitation wavelengths employed for the preparation of the 5_1 and 3_1 and 4_1 states. Therefore, the increase in the intensity of the center of the arrival-time distribution is attributed to an increase in production of Cl and Cl* photofragments with nearly zero center-of-mass translational energies.

The β parameters and the velocity distributions of the photofragments were extracted by simulating the TOF spectral profiles.[34] The distributions of both Cl and Cl* resulting from the VMP of 3_1 and 4_1 are significantly narrower than that of 5_1. It was also observed that the broadening of the latter results from production of "faster" Cl and Cl* photofragments as well as slower ones. The production of faster photofragments arises from higher combined energies (IR/ VIS + UV) employed in the VMP of CHFCl$_2$ 5_1 ($56\,459\,\mathrm{cm}^{-1} \cong 7.0\,\mathrm{eV}$) than in 4_1 ($53\,873\,\mathrm{cm}^{-1} \cong 6.7\,\mathrm{eV}$) and 3_1 ($51\,236\,\mathrm{cm}^{-1} \cong 6.4\,\mathrm{eV}$). The combined energies in the VMP of CHFCl$_2$ 3_1, 4_1 and 5_1 exceed that required for the loss of one chlorine atom, but the first two do not surpass the threshold for a three-body process where two ground-state chlorine atoms are released. Therefore, it seems likely that the slower Cl and Cl* photofragments, observed in the VMP of CHFCl$_2$ 5_1, emerge from the three-body decay, either synchronous concerted, where two C–Cl bonds are broken simultaneously, or sequentially.[48,49]

The recoil anisotropies also provide some information regarding the mechanism of bond breaking. It was seen that the Cl and Cl* arising from ~ 235-nm photodissociation of CHFCl$_2$ 3_1, 4_1 and 5_1 *via* two- or three-body processes possess positive anisotropies, lower than the limiting values. The β parameters of Cl increase somewhat with increasing combined energy, rising from $\beta = 0.14 \pm 0.05$ to 0.27 ± 0.06 and to 0.47 ± 07, while those of Cl* are nearly constant with values of $\beta = 0.36 \pm 0.06$, 0.43 ± 0.05 and 0.34 ± 0.04. The theoretical limit for the β parameter, with μ parallel to the line connecting the two Cl atoms, was calculated and estimated to be ~ 1.1 (based on a Cl–C–Cl bond angle of $112°$ in CHFCl$_2$).[23] In 193-nm photodissociation of vibrationless ground-state CHFCl$_2$ molecules, $\beta = 0.5 \pm 0.1$ was measured, indicating a partial loss of anisotropy.[23] This loss was attributed either to a geometrical rearrangement during dissociation (nonaxial recoil) or to a small contribution of a perpendicular electronic $A' \leftarrow A'$ transition overlapping the dominant $A'' \leftarrow A'$ parallel transition.

The β values obtained in the VMP experiment are also positive, presumably due to the access to similar upper electronic states as in the (energetically close) 193-nm photodissociation, *i.e.*, involvement of the A' and A'' transitions in the absorption from the vibrationally excited state. Indeed, calculations[50] have shown that two transitions of A'' and two of A' symmetry underlie the first absorption band of CHFCl$_2$, related to the $\sigma^* \leftarrow n$ transition, and that the predominant contribution rises from the lowest-energy A'' state and some from one of the A'. Relying on the accessibility of the upper dissociative states it is conceivable that the dissociation is prompt and it does not seem that the rotational motion accounts for the reduction of β from its limiting value. Thus, the observation of less than limiting β values emerges from dynamical factors. From the measured β it is inferred that both Cl and Cl* are produced as a result of simultaneous absorption to both A'' and A' states that mix, probably, *via* curve crossing to release Cl and Cl*. Also, due to the increase of the β corresponding to Cl, it is likely that the contribution of the A'' state to this channel increases with increasing combined energy.

The observed increase in the Cl channel anisotropy in VMP of $CHFCl_2$ molecules promoted from 5_1 relative to those from 4_1 and 3_1 is a hint for a concerted three-body decay. This is because in the case of sequential three-body decay one would expect a lower β parameter than in the two-body case due to the increased dissociation lifetime. Relying on the observation of the increase in β it seems likely that the concerted three-body decay is responsible for the slow Cl. As for Cl* photofragments, due to the small decrease in β in VMP of $CHFCl_2$ 5_1 relative to that in 4_1, it might be that a sequential three-body decay is also involved. These results suggest that the photodissociation of $CHFCl_2$ 3_1 and 4_1 occurs only *via* two-body decay, while that of $CHFCl_2$ 5_1 occurs *via* three-body decay as well. Not surprisingly, it appears that the onset of the three-body decay is observed when the combined energy of the vibrational excitation and photodissociating photon overcomes the dissociation barrier for it.

8.2.1.2 CH_3CFCl_2

(a) **Enhanced Production of Photofragments.**   VMP of CH_3CFCl_2 was studied for molecules prepared in the fundamental CH_3 symmetric stretch, $1\nu_{CH}$ (by SRE)[31], and in the second, $3\nu_{CH}$,[32,36] third, $4\nu_{CH}$,[32,36] and fourth, $4\nu_{CH}$,[32,35] CH_3 overtones.

The signal intensities obtained in the VMP of $3\nu_{CH}$ and $4\nu_{CH}$ of CH3CFCl$_2$ relative to those obtained in $\sim 235/243$-nm photodissociation of the vibrationless ground state reveal that the production of the atomic photofragments is significantly enhanced as a result of vibrational excitation. This enhancement can be due to the combined photon energies that are larger by ~ 8580 and ~ 11 $270\,cm^{-1}$ for $3\nu_{CH}$ and $4\nu_{CH}$, respectively, but also due to better FC factors. Moreover, the observed enhancement in the photodissociation stage is larger for $4\nu_{CH}$ than for $3\nu_{CH}$, possibly due to both the increase in energy and to a better FC overlap with the upper PESs. However, as for $CHFCl_2$, although C–H methyl stretches are excited, not only C–H bond breaking is enhanced, but large enhancements are obtained for both Cl and Cl* photofragments indicating that the energy flows into the C–Cl bond and leads to its effective cleavage.

(b) **The Effect of Vibrational Pre-Excitation on the Branching Ratios of the Atomic Photofragments in the VMP of CH_3CFCl_2 and CH_3CF_2Cl.**   Table 8.1 compares the Cl*/Cl and H/(Cl + Cl*) branching ratios in photodissociation of vibrationless ground state to those in VMP of CH_3CFCl_2[32] and CH_3CF_2Cl.[38] The Cl*/Cl branching ratios for $3\nu_{CH}$ and $4\nu_{CH}$ VMP are comparable. However, they are about twice that for 193-nm photodissociation although the total excitation energies in all three photodissociation processes are close. Moreover, the energy of the 193-nm photodissociation is in-between the combined energies employed in the $3\nu_{CH}$ and $4\nu_{CH}$ VMP, which

Table 8.1 Cl*/Cl and H/(Cl + Cl*) branching ratios in direct photodissociation of the vibrationless ground state and in VMP of CH_3CFCl_2[32]
and CH_3CF_2Cl.[38]

	Direct photodissociation			VMP	
CH_3CFCl_2	193 nm	235 nm	$1\nu_{CH}$	$3\nu_{CH}$	$4\nu_{CH}$
Energy (cm^{-1})^a	51 813	42 492	45 448	51 071	53 761
Cl*/Cl	0.22 ± 0.06	0.36 ± 0.08	0.54 ± 0.11	0.49 ± 0.11	0.46 ± 0.10
H/(Cl + Cl*)^b	0.17 ± 0.07	0.36 ± 0.12	0.46 ± 0.15	0.23 ± 0.09	0.32 ± 0.12
CH_3CF_2Cl	193 nm			$3\nu_{CH}$	$4\nu_{CH}$
Energy (cm^{-1})^a	51 813			∼ 51 150	∼ 53 850
Cl*/Cl	0.25 ± 0.06			0.54 ± 0.09	0.55 ± 0.09

^{a}The energy given in the table for VMP represents the combined energy for vibrational excitation + photodissociation. The energy for the photodissociation part in the VMP was taken as that
used when Cl was tagged, 235 nm.
^{b}The combined energy in the case of H tagging is 1363 cm^{-1} lower than for Cl, which might affect
the determined H/(Cl + Cl*) ratios.

leads to the conclusion that the vibrational pre-excitation itself causes the
increase in the branching ratio. Support for this conclusion can be obtained
from the H/(Cl + Cl*) branching ratio in CH_3CFCl_2 that also seems to
increase upon pre-excitation. The effect of vibrational pre-excitation is also
exemplified in CH_3CFCl_2 by comparing 235-nm direct photodissociation
with VMP of the fundamental CH_3 symmetric stretch, $1\nu_{CH}$ (although the
excitation energy in the former is lower): both Cl*/Cl and H/(Cl + Cl*)
branching ratios increase upon pre-excitation. It is noteworthy that the
branching ratios for 193 nm *are lower* than for 235 nm although the excitation energy in the former is higher.

The effect of vibrational pre-excitation on branching ratios was experimentally and theoretically studied in VMP of other haloalkanes (Sections 8.2.2 and
8.2.3). Those studies indicate that initial vibrational excitation of the alkyl
group alters the branching ratio between the spin-orbit states of the ensuing
halogen by altering the photodissociation dynamics, often *via* curve crossing. It
is also possible that the increased Cl*/Cl ratio in VMP of CH_3CFCl_2 and
CH_3CF_2Cl is due to both their direct promotion to A' states and to curve
crossing to these states that correlate with Cl*. The increase of H/(Cl + Cl*)
branching ratios in the VMP of CH_3CFCl_2 may be related to the initial
extension of the C–H bonds upon the vibrational pre-excitation.

8.2.2 Methyl Iodide

8.2.2.1 Photodissociation in the First Absorption Band: General Background

There has been much interest in photodissociation of methyl iodide, CH_3I, in
the first absorption band (the *A*-band, ∼225–335 nm) due to its benchmark
role in studying multichannel dissociation of simple polyatomic molecules.[51–61]

This is since the symmetric top CH_3I fragmentation to I atoms occurs along the C_{3v} symmetry axis, providing a better model of the linear triatomic system dissociation than real linear triatomics do because of the strong bending vibration influence in excited states of the latter molecules. Also, CH_3I is a source of excited iodine atoms, I*, in the active medium of photodissociation iodine lasers. As shown below, vibrational excitation can control the production of I*, hence the special interest in VMP of CH_3I.

The literature on the UV photodissociation of CH_3I in its first absorption band is very extensive (Ref. 60, published in 2007, lists 82 papers on UV photodissociation of CH_3I in its first absorption band and states that the list is far from being complete). In References 51–61 we list only papers dealing explicitly with VMP or those quoted in the following discussion on VMP of CH_3I. Although updated calculated potential-energy curves for the low-lying states of CH_3I are presented in Ref. 60, we show here a simpler version[58] that contains the basic information needed for our discussion (Figure 8.5). Forming the A-band, there are three electronic states associated with the excitation,

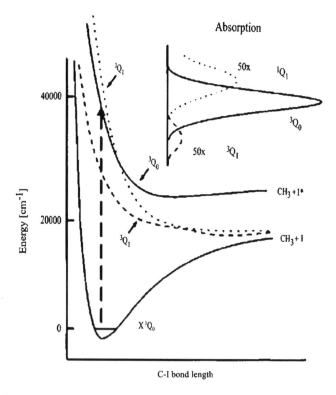

Figure 8.5   Schematic diagram of the relevant PESs and their dissociation limits for the A-band of CH_3I. In the upper corner the decomposed A-band absorption spectrum is shown. Reprinted with permission from Ref. 58. Copyright 1999, American Institute of Physics.

namely $^3Q_1$, $^3Q_0$ and $^1Q_1$ in Mulliken's notation,[62] which can be optically accessed from the ground electronic state, $\tilde{X}^1Q_0$. These states correspond to $a\ ^3\Pi_1$, $a\ ^3\Pi_0$, $A\ ^1\Pi_1$ and $X\ ^1\sum^+$, respectively, in the related diatomic HI molecule.

The photophysics of CH_3I relevant to VMP is summarized as follows:[58] The *A*-band is dominated by a parallel transition to the $^3Q_0$ state that correlates adiabatically with the products $CH_3 + I^*$. Perpendicular transitions to the $^1Q_1$ and $^3Q_1$ states, which correlate with ground-state iodine atoms, I, are also allowed optically, but are much weaker, <2% of the total absorptivity. The experimental observation of a large yield of I atoms with an angular distribution characteristic of a parallel transition implies that curve crossing takes place, *i.e.*, crossing at the seam of the $^3Q_0$–$^1Q_1$ CI. Off-axis nuclear motion that presumably arises from the zero-point motion of the degenerate (*e*) vibrational modes of CH_3I couples the two surfaces *via* the kinetic energy operator. The balance between the parallel and perpendicular character of this (I) channel is a direct probe of the contributions from the $^3Q_0$ and $^1Q_1$ states, respectively. Photodissociation on the $^3Q_0$ surface is extremely rapid and axial, and the CH_3I retains C_{3v} symmetry during the initial stages of dissociation. The rapid traverse from a bound molecule to separated fragments, reflecting the steep walls of the repulsive electronic states, is confirmed by direct measurement of an appearance time of $\sim 150\,fs$ for the fragments using ultrafast pump–probe laser techniques. For the parallel $^3Q_0 \leftarrow \tilde{X}$ transition, the transition dipole points along the C–I bond, thus molecules with bond axes parallel to the **E** field direction of the linearly polarized light beam are primarily excited and dissociated along the laser polarization direction. The anisotropy approaches the maximal value possible $\beta \cong 2$ for the I* channel.

8.2.2.2 *Photodissociation of Vibrationally Excited CH₃I*

Comparison of the geometric structure and normal vibrational modes of methyl iodide and the methyl radical is instrumental in anticipating vibrational energy disposal in photodissociation of vibrationally excited CH_3I.[53] Figure 8.6 shows the energies and nuclear motions for NMs in C_{3v} symmetry for CH_3I and CH_3. Three CH_3I modes, $v_1(a_1)$, $v_4(e)$, and $v_5(e)$ appear in the CH_3 radical as $v_1(a_1)$, $v_3(e)$, and $v_4(e)$, respectively, with essentially the same relative motion and energy. The energy of the umbrella mode $v_2(a_1)$ of CH_3I (1254 cm^{-1}) drops by half in the equivalent $v_2(a_1)$ mode of the CH_3 radical (606 cm^{-1}) on removal of the I atom. Two CH_3I modes disappear, $v_3(a_1)$, the C–I stretch, which couples directly into the dissociation coordinate, and $v_6(e)$, the degenerate methyl rock. Based on this very qualitative picture alone, (ignoring, *e.g.*, the topological features of the PESs) it can be anticipated that vibrational energy in the CH_3I modes, $v_1(a_1)$, $v_4(e)$, and $v_5(e)$, would appear as vibrational energy in the CH_3 product, while energy in $v_3(a_1)$, the C–I stretch, would appear as increased translational energy. $v_6(e)$, the degenerate methyl rock, should be an important coupling mode for promoting curve crossing, and v_6 internal energy could appear in the methyl fragment as transverse recoil in the form of increased out-of-plane rotation or as excess translational energy.

Normal Modes of CH₃I and CH₃

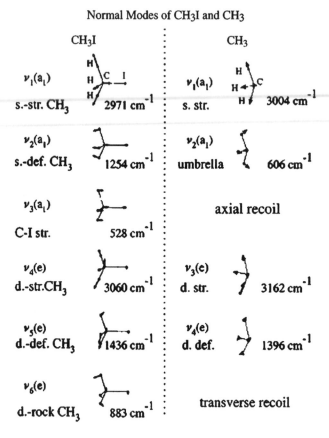

Figure 8.6 Energy and symmetry (C_{3v} nomenclature) of the NMs of CH_3I and the CH_3 radicals. The relative nuclear motions are shown schematically. Reprinted with permission from Ref. 58. Copyright 1999, American Institute of Physics.

In the VMP experiments described below, photodissociation at different wavelengths across the A-band of a molecular beam of CH_3I was carried out.[58] The source of vibrational population of the ground electronic state of CH_3I was thermal. State-selective detection of the I, I*, and $CH_3(v)$ photoproducts was performed using VMI. At room temperature, the Boltzmann population of the lower-energy CH_3I vibrational modes is $v_3(a_1) \sim 8\%$ and the doubly degenerate $v_6(e) \sim 5\%$. Vibrational cooling in a supersonic expansion is much less efficient than rotational cooling, especially for higher-energy vibrations. Since, to preclude signal from clusters, the early stages of the molecular-beam expansion was probed, a significant portion of the initial room-temperature population of vibrationally excited molecules was encountered.

In the measured photofragments kinetic-energy release curves obtained in the photodissociation of CH_3I,[58] an extra peak for $CH_3(v=0)$ in both I and I* channels was observed at higher kinetic-energy release. This is indicative of molecules excited in a CH_3I vibrational mode that couples into the dissociation coordinate, most likely v_3 (C–I stretch) of CH_3I. A Lorentzian fit to the angular distributions for the I channel $CH_3(v=0)$ peak suggested that two contributions (peaks) were present, one from cold CH_3I, yielding $v=0$ CH_3 with $\beta=1.7$, and the second a channel $\sim 900\,cm^{-1}$ higher in energy with a slightly negative beta parameter, ~ -0.2. An extra energy release of $\sim 900\,cm^{-1}$ corresponds very well with the dissociation of CH_3I excited in $v_6=1$ ($883\,cm^{-1}$), the degenerate CH_3 rocking mode. This peak-to-peak spacing also implies that the v_6 vibrational energy is fully converted into translational energy release.

From a careful analysis of the above-mentioned and of additional experimental results,[58] the role of the $v_3(a_1)$ symmetric C–I stretch and the $v_6(e)$ degenerate methyl rock in the photodissociation dynamics CH_3I can be summarized as follows. The C–I stretch excitation appears as excess translational energy exclusively in the I* channel, with an angular distribution identical to that of the ground-state CH_3I molecules very near the maximum *parallel* anisotropy ($\beta \sim 1.9$). A totally symmetric vibration does not appear to enhance curve crossing or change the relative absorption strengths of the $^3Q_0$ to $^3Q_1$ states. The absolute absorption strength to the $^3Q_0$ state was greatly enhanced, however, far beyond the expected population of these levels at room temperature. Moreover, even the $v_3=2$ state of CH_3I was observed, although it has a Boltzmann population at room temperature of less than 1%. FC overlap is clearly much stronger for the higher v_3 levels, as calculated in Refs. 55 and 57. From Ref. 55 an enhancement of at least a factor of 20 for $v_3=1$ and 100 for $v_3=2$ can be expected.

It was found that CH_3I in the v_6 mode leads to CH_3 fragments in the $v=0$ state, primarily in the I channel, and a *perpendicular* angular distribution.[58] The channel specificity of the two CH_3I vibrational modes is quite strong. There are many possible explanations for the behavior of the $v_6(e)$ mode. One possibility is that the $^3Q_0$ state remains the upper state in the optical excitation but that the transition becomes perpendicular, *i.e.*, $^3Q_0(a) \leftarrow \tilde{X}(e)$ is a perpendicular transition, with the fragments ejected preferentially in the horizontal plane. An *(e)* symmetry mode is expected to greatly enhance curve crossing, thus the production of I atoms. It might then be expected that the angular distribution for the v_6 contribution would be $\beta=-1$. The analysis of the results suggests a lower value, $\beta=-0.2$. This could be consistent with a partial contribution ($\sim 25\%$) from the $^3Q_1$ state, *i.e.*, $^3Q_1(e) \leftarrow \tilde{X}(e)$, which would thus give a parallel fragment angular distribution (correlating directly with the I channel), raising the overall beta from -1 to -0.2.

To conclude the section on methyl iodide we mention additional points inferred in theoretical works that are relevant to the interpretation of the VMP of this molecule. Comparing the influence of bending in CH_3I *vs.* CD_3I it was found that parent bending motion (*e.g.*, $v_6(e)$, but any e-symmetry vibration contains bending) has a significant effect on the final electronic branching of dissociation fragments.[56] The calculation generated a larger I* yield from the CD_3I dissociation than that from CH_3I, The oscillator strength is largely carried by the $^3Q_0$ state that diabatically correlates with the I* channel. Thus, the fragment I is largely produced *via* nonadiabatic transitions from the $^3Q_0$ state to the $^1Q_1$, state. A smaller amplitude in the bending coordinate in CD_3I, as compared to CH_3I, results in a weaker interaction between two diabatic surfaces, and leads to a larger I* yield. These results are in line with the conclusions of the work on VMP of CF_3I presented below in Section 8.2.3.1.

A comprehensive *ab initio* study of CH_3I photodissociation, including vibrational state control of the I* quantum yield, is presented in Ref. 61. The computational method employed was based on a combination of the multireference configuration interaction approach and relativistic effective core potentials. Dipole and transition moments for the ground and low-lying excited states of CH_3I, and on this basis partial and total absorption spectrum of the *A*-band and the I* quantum yield as a function of excitation energy were calculated.

The calculated $^3Q_0$, $^3Q_1$, and $^1Q_1 \leftarrow \tilde{X}$ transition moments show a strong dependence on the C–I distance in the FC region, which is opposite for the parallel and perpendicular transitions. This makes the assumption of constant values for these quantities, which was previously employed for obtaining the corresponding potentials from experimental data and modeling the *A*-band partial contributions, a poor approximation. On the basis of the calculated absorption data, the I* quantum yield was determined as a function of excitation energy. It is shown that the strong dependence of the relevant transition moments on the C–I distance opens up a possibility for vibrational state control of the final photodissociation products. It is demonstrated that I* quantum yields up to 0.9 can be achieved when vibrationally hot CH_3I molecules, $v = 1,2$ of the C–I stretch, v_3 528 cm^{-1} mode (the v_6 883 cm^{-1} mode or other modes were not included in the calculations) are excited in the 32–35×10^3 cm^{-1} spectral range. This effect also reveals itself in a strong temperature dependence of the I* quantum yield values, particularly in the low-energy part of the *A*-band. It is emphasized that vibrational state control of the I*/I branching ratio in CH_3I photodissociation has an electronic rather than a dynamic nature. Due to a different electron density distribution at various molecular geometries, one achieves more efficient excitation of a particular fragmentation channel rather than influences the dynamics of the decay process. The results presented in this work[61] imply that photoexcitation of the vibrationally hot CH_3I molecules in the above energy range provides a good possibility for obtaining more efficient laser generation at the atomic I* $\rightarrow$ I transition.

It is worth noting that complete, coherent control of the photoproducts in multipath two-photon dissociation of CH_3I was computationally demonstrated.[59,59a] The v_2 and v_3 vibrational levels in the ground electronic state were used, respectively, as intermediates in the two-photon dissociation pathways. The control is achieved by interference of the quantum pathways on the competing channels.

8.2.3   Other Haloalkanes

In pioneering experiments on VMP of haloalkanes, the v_1 mode of CF_3I was excited by a CO_2 laser at 9.6 μm[63] and the v_6 mode of CH_3Br by a CO_2 laser at

$10.6 \, \mu m$.[64] In both cases it was concluded that vibrational pre-excitation enhances the photodissociation. In later VMP experiments on CF_3I[65] (as on CH_3I, Section 8.2.2.2 above), low-lying vibrational states were thermally populated. In the other experiments,[26,66–73] pre-excitation in the region of the second, third and fourth CH overtones was carried out using IR/VIS lasers.

Of the vast amount of work carried out on VMP of the "other haloalkanes",[26,63–74] we dwell here, mainly, on the branching ratio of the atomic photoproducts.[26,65,69–74]

8.2.3.1 CF_3I

Photodissociation of vibrationally excited CF_3I (prepared by heating the sample) was found to increase the branching to ground-state I products.[65] The experiment tested a model for the dependence of I/I* branching at a CI on the amplitude of the dissociative wavefunction at bent geometries. An increase was observed in the branching from 13% to 17% I photoproducts when the temperature of the CF_3I parent was increased from 100 to 400 °C, in agreement with the qualitative prediction of the model. The experiment demonstrated the importance of the initial vibrational eigenstate in moderating nonadiabatic effects in molecular photodissociation. These findings are in line with the above results on comparing the influence of bending in CH_3I *vs.* CD_3I[56] quoted in Section 8.2.2.2.

8.2.3.2 CH_3Cl and CHD_2Cl

An increase in the branching into Cl* was observed in VMP of CH_3Cl[26,69,70] and CHD_2Cl[70] pre-excited to the fourth overtone of the C–H stretch, with a larger increase in the former,[70] Also, the total yield of atomic Cl fragments from the VMP of CH_3Cl was significantly higher than for CHD_2Cl, more than can be accounted for by differences in the cross sections for overtone excitation. This suggests that state mixing leads to greater amplitude of the wavefunction of the vibrationally excited level along the dissociation coordinate in CH_3Cl than in CHD_2Cl. The observed differences in the Cl spin-orbit branching in the photolysis of ground state *vs.* vibrationally excited molecules were ascribed to differences in the nuclear dissociation dynamics, rather than access to other excited electronic states.[70] An increase in Cl* was also observed in CH_3Cl pre-excited to the second and third overtones.[71]

A slight preferential formation of D over H atoms in the 243-nm VMP of CHD_2Cl ($v_{CH} = 5$) was found.[70] This result is quite interesting, in view of the observed preference for H atom formation in the photolysis of ground-state CHD_2Cl. The excited vibrational level is accessed through the oscillator strength associated with the fourth C–H stretch overtone transition. Since for $v_{CH} = 5$ there is strong coupling of C–H and C–D stretch excitation,[75,76] it would be very interesting to determine the ratio of the H to D atom yields in the VMP of other overtone excited levels of CHD_2Cl. In particular, vibrational energy levels associated with lower C–H stretch overtone excitation ($v_{CH} \leq 4$) are predicted to have considerably less C–D stretch excitation character, because of the detuning of the zero-order C–H and C–D anharmonic vibrational levels with decreasing v_{CH}.[70]

8.2.3.3   *CH₂Cl₂*

The intramolecular dynamics in the VMP of dichloromethane was studied *via* ~235-nm photodissociation of pre-excited CH_2Cl_2 in the regions of the second,[72,73] and of the third and fourth[73] overtone stretches ($v_{CH} = 3\text{–}5$). The action and PA spectra exhibit a multiple-peak structure corresponding in terms of NM and LM models to C–H stretch overtones and combinations of C–H stretches and bends. The splittings between the features reflect subpicosecond redistribution times between the stretch-bend states, which are between those encountered for molecules with isolated C–H and methyl groups. The energy flow from the C–H to the C–Cl bond is evident from the increased ground and spin-orbit-excited chlorine yield. The Cl*/Cl branching ratio in the VMP is higher than that obtained in the direct, almost isoenergetic, 193-nm photodissociation of CH_2Cl_2 and is suggestive of dynamics proceeding more nonadiabatically in the former. The nonadiabaticity of the process is also supported by the measured anisotropy parameters. These parameters are lower than the limiting values of a pure $A'' \leftarrow A'$ parallel transition, indicating involvement of A' states, which nonadiabatically interact to alter the Cl*/Cl ratio in the VMP.

It was suggested[73] that although ground-state CH_2Cl_2 is of C_{2v} symmetry, the initial CH vibrational excitation and the energy redistribution impact the geometry, distorting it and reducing its symmetry to C_s in the FC region. This facilitates transitions from the vibrationally excited state of the A' ground state to repulsive states where A' and A'' states interact. This conclusion is supported by spin-orbit *ab initio* calculations on the photodissociation of CH_2Cl_2.[74]

8.3   VMP of Phenol

Phenol is the chromophore of the aromatic amino acid tyrosine and its photoinduced processes have emerged as a prototype for the exploration of the electronic spectroscopy and photochemistry of aromatic biomolecules. A time-dependent quantum wavepacket description of the photochemistry of phenol, including VMP, is presented in Ref. 77 and general background to the photodissociation of phenol, including from hot bands, is given in Ref. 78. The PESs of phenol relevant to VMP studies[79] are presented in Figure 8.7. Electronic-structure calculations, summarized in Ref. 77, revealed that the UV photochemistry of phenol essentially involves the three lowest electronic states S_0, $^1\pi\pi^*$, and $^1\pi\sigma^*$. The first excited state, corresponding to excitation from the highest-occupied molecular orbital (HOMO) of π character to the lowest unoccupied molecular orbital (LUMO) of π^* character, is the origin of the

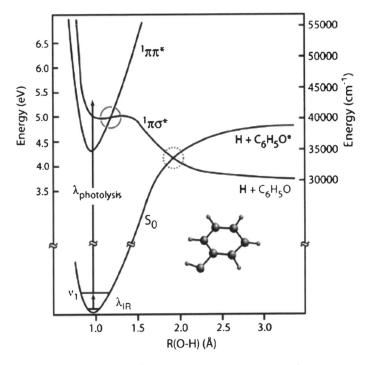

Figure 8.7 PESs of phenol. The $^1\pi\pi^*$ state origin lies $36\,350\,\text{cm}^{-1}$ ($4.507\,\text{eV}$) above the ground state. The $^1\pi\sigma^*$ state forms a CI (marked by a solid circle) with the $^1\pi\pi^*$ state at about $40\,300\,\text{cm}^{-1}$ ($5.0\,\text{eV}$). The second CI (marked by a dotted circle) is between the $^1\pi\sigma^*$ state and the ground state. The dissociation energy to the ground state is $30\,015 \pm 40\,\text{cm}^{-1}$ and the excited phenoxyl state energy is $8900\,\text{cm}^{-1}$ above the ground state. The arrows show the vibrational and electronic excitation in the VMP experiments. Reprinted with permission from Ref. 79. Copyright 2008, American Institute of Physics.

absorption band around 4.5 eV. The second excited state corresponds to the excitation to the Rydberg type 3s orbital and is optically dark. This 3s Rydberg orbital has a significant antibonding σ^* character with respect to the OH bond, opening a pathway for hydrogen abstraction. Along this coordinate, the $^1\pi\sigma^*$ state intersects the other two states, resulting in two curve crossings that become CIs when out-of-plane modes are taken into account. Due to the existence of a CI between the bright $^1\pi\pi^*$ state and the dark $^1\pi\sigma^*$ state, the population of the bright state can be transferred to the dark state through an IC after excitation.

The presentation below of the VMP experiments on phenol follows Ref. 79. In these experiments an IR photon was used to excite the O–H stretching state (v_1) of partially deuterated phenol (phenol-d_5, C_6D_5OH) and a UV photon to transfer it to the $^1\pi\pi^*$ electronically excited state. (3 + 1) REMPI, in combination with VMI detection, monitored the H atoms produced following the electronic excitation. The H-atom velocity distributions were converted to a recoil-energy distribution. Electronic action spectra were obtained by monitoring the H atoms from the photodissociation of phenol-d_5 while varying the wavelength of the electronic excitation laser. Phenol-d_5 was used to suppress the large background H-ion signal that comes from $(1+1)$ REMPI of phenol by the photolysis pulse and dissociation of the resulting cations by the probe pulse. Deuterating the C–H bonds eliminated the part of the background signal that comes from the hydrogen atoms on the phenyl ring.

The total recoil-energy distributions from VMP and one-photon photodissociation of phenol-d_5 are shown in Figure 8.8. The most striking result is that dissociation of phenol-d_5 molecules with an initially excited O–H stretching vibration produces significantly more fragments with low recoil energies than does one-photon dissociation at the same total energy, in particular for the higher excitation energies. The intensity of the prominent high-energy feature at $12\,000\,cm^{-1}$ decreases in favor of a lower-energy feature, an observation that is consistent with increased production of electronically excited phenoxyl-d_5 fragments.

These results were rationalized[79] in accordance with the above-mentioned theoretical predictions.[77] The enhanced yield of low recoil energy fragments compared to one-photon dissociation suggests that initial excitation of the O–H stretch promotes the formation of excited-state phenoxyl fragments. At one-photon excitation energies above the $^1\pi\pi^*-^1\pi\sigma^*$ CI, phenol molecules can pass through the intersection and dissociate directly on the $^1\pi\sigma^*$ state to make high recoil energy $(12\,000\,cm^{-1})$ fragments. In addition, some of them internally convert and decay statistically on the ground electronic state with low recoil energy. Excitation from an initially excited O–H stretching state to the same total energy changes the situation substantially to produce more low recoil energy fragments. These fragments are mostly excited-state phenoxyl radicals resulting from passage around the $^1\pi\sigma^*-S_0$ CI and dissociation on the excited-state potential. The situation is more complicated at lower total energies, where the excitation is above the $^1\pi\pi^*-^1\pi\sigma^*$ CI only with the inclusion of the O–H stretching excitation. Because energy tied up in O–H stretching does not promote passage through this CI, IC, with or without preservation of O–H stretching excitation, becomes more important. This competing pathway decreases both the direct dissociation to produce ground-state phenoxyl fragments with large recoil energies and the passage around the $^1\pi\sigma^*-S_0$ CI to produce excited-state phenoxyl fragments. The inference of enhanced production of excited phenoxyl fragments from phenol molecules with an excited O–H stretching vibration at higher total energies agrees qualitatively with the predictions of the theoretical work[77] that O–H stretching excitation promotes the adiabatic decomposition.

8.4   VMP of Other Larger than Tetratomic Molecules

In the previous sections of this chapter we presented examples where a variety of features of VMP were observed and interpreted. Most of these features were observed also in other "larger than tetratomic" molecules and the reader of the

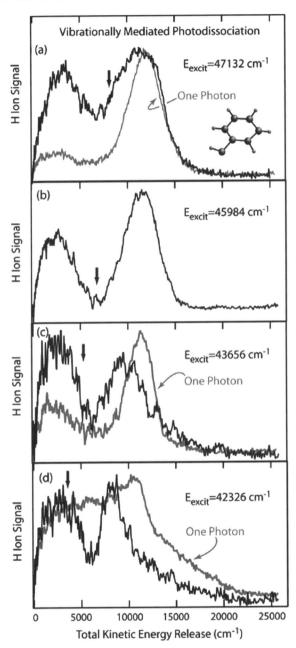

Figure 8.8 (a) Total recoil-energy distributions from VMP (dark trace) and one-photon (light trace) photodissociation of phenol-d_5 at a total excitation energy of $E_{excit} = 47\,132\,cm^{-1}$. (b) Distribution from VMP at $E_{excit} = 45\,984\,cm^{-1}$. (c) Distributions for $E_{excit} = 43\,656\,cm^{-1}$. (d) Distributions for $E_{excit} = 42\,326\,cm^{-1}$. The vertical arrows mark the maximum recoil energies possible for excited-state phenoxyl, (a) 8230, (b) 7080, (c) 4756, and (d) 3430 cm^{-1}. Reprinted with permission from Ref. 79. Copyright 2008, American Institute of Physics.

previous sections should be able to follow the rather brief description and interpretation of the VMP of these molecules, presented below. For convenience we present these "other" molecules in three groups. The first group, Section 8.4.1, deals with molecules containing OH, where, mostly, O–H stretches were initially excited.[80-88] The second and third group deal with molecules where C–H stretches (or combination of these stretches and other modes) were initially excited, acetylene homologues[17,36,89-98] in Section 8.4.2 and ethene isotopologues[99-101] in Section 8.4.3. In all groups the VMP dynamics was studied, but in the last two a significant part of the work was devoted to application of the VMP to vibrational spectroscopy, including assignment of vibrational states. In addition to these three groups we very briefly describe, in Section 8.4.4, VMP related work on UF_6.

8.4.1 $HONO_2$, $(CH_3)_3COOH$, CH_3OH, NH_2OH, HO_2NO_2 and CH_3OOH

8.4.1.1 $HONO_2$ ns and fs-ps VMP Studies

In one of the earliest detailed VMP studies of a relatively large molecule, nitric acid, $HONO_2$, was excited in the region of the third overtone of the O–H stretching vibration ($4\nu_{OH}$).[80,81,83] The OH fragment, detected by LIF, determined the energy partitioning and identified the influence of vibrational excitation prior to dissociation. VMP using 755- plus 355-nm photons deposited more energy in relative translation than the isoenergetic one-photon dissociation using 241-nm photons. The former process also produced three times more vibrationally excited OH fragments, and both processes formed electronically excited NO_2, which received over three-quarters of the available energy. The latter observation is probably due to the fact that the electronic excitation is primarily localized on the NO_2 chromophore. In addition, it was found that vibrational overtone excitation enhanced the cross section for the subsequent electronic excitation by about three orders of magnitude. The observed differences are consistent with the motion of the vibrationally excited molecule on the ground electronic state surface strongly influencing the dissociation dynamics by allowing access to different electronic states in the photolysis step.

VMP experiments on $HONO_2$ were also carried out in the fs-ps time domain.[85] The first overtone of the O–H stretch vibration in $HONO_2$ was excited with a 100-fs laser pulse. A second, time-delayed pulse preferentially photodissociated molecules having vibrational excitation in modes orthogonal to the O–H stretch. The photodissociation yield increased as a function of time because energy flows out of the initially excited O–H bond into other more efficiently dissociated vibrations. A single exponential was observed for this intramolecular vibrational relaxation with a time constant of 12 ps. This is consistent with moderate coupling of the O–H stretch to states close in energy.

8.4.1.2 $(CH_3)_3COOH$

The VMP of *tert*-butyl hydroperoxide (TBH), $(CH_3)_3COOH$, is similar to that of HOOH in the sense that both molecules contain O–O bond that is broken upon photodissociation. As in some of the studies of HOOH (Section 7.4.1), in the VMP study of TBH the $4\nu_{OH}$ vibrational level was excited with a $\sim 13\,300\,cm^{-1}$ photon and a second photon with the same energy dissociated the molecule.[81,82] Comparison of the energy disposal in the VMP with that for the isoenergetic one-photon dissociation at 376 nm, illustrated the unique dynamics that can occur when vibrational excitation precedes photodissociation in this large molecule. Single-photon photolysis produced fragments with large recoil velocities, while VMP produced slowly recoiling fragments having substantially more energy in internal excitation.

The OH products have less rotational energy than those obtained from the corresponding process in HOOH. Also, less energy is released into translation for TBH than for HOOH and $HONO_2$ (previous section) and no evidence for the formation of $OH(X\,^2\Pi, v=1)$ was found. This again contrasts with the results for HOOH and $HONO_2$ in which $\sim 11\%$ and $\sim 5\%$ of the OH fragments are formed vibrationally excited in the VMP process, respectively. The total energy deposited in the parent molecule in the TBH experiments is $\sim 26\,600\,cm^{-1}$, which exceeds the threshold for O–O bond fission by $\sim 12\,300\,cm^{-1}$. This excess energy is partitioned among the translational, rotational and vibrational degrees of freedom of the $(CH_3)_3CO$ and OH photofragments, but very little of this excess energy is retained in the OH fragment. This is consistent with the t-butyl group in TBH providing an energy "sink" that retains much of the excess energy in the numerous internal quantum states of the $(CH_3)_3CO$ fragments.

An important issue is the extent of O–H and C–H vibrational mode mixing in TBH. Initial C–H overtone excitation might be expected to be less efficient as compared to O–H overtone excitation in promoting further photon absorption and subsequent dissociation. However, the weakness of the C–H overtone transitions in this region did not enable, thus far, this issue to be addressed properly.

8.4.1.3 CH_3OH

Unlike the other molecules discussed in this section, but like water, theoretical studies[84] of VMP of methanol, CH_3OH, preceded the experimental studies.[87] In the theoretical work, photodissociation of CH_3OH in the first absorption band ($S_1 \leftarrow S_0$) was investigated by a two-dimensional wavepacket study employing the associated internuclear PESs obtained from *ab initio* calculations. The emphasis in this work was on exploring the effect of excitation of the O–CH_3 mode, prior to excitation with the photolysis photon, on the photodissociation channels. The branching ratio for the $CH_3 + OH$ and $CH_3O + H$ channels was shown to exhibit a drastic dependence on the initial vibrational state of CH_3OH as well as on the energy of the dissociating photon. Whereas dissociation of the vibrational ground state yields exclusively the $CH_3O + H$ products, increasing the excitation of the $O–CH_3$ mode access more and more of the $CH_3 + OH$ channel. However, excitation to levels with at least five or six quanta in the $O–CH_3$ mode are required to obtain a reasonably large yield of OH radicals. It was suggested that a proper selection of the initial vibrational state and of the dissociating photon frequency should enable control of products to be exerted to obtain branching ratios spanning several orders of magnitude.

In the experimental work,[87] preparation of the O–H fundamental stretch of jet-cooled CH_3OH was followed by its excitation to the S_1 surface and the ensuing H photoproducts were monitored by $2 + 1$ REMPI. It was found that the maximum in the electronic excitation action spectrum of the vibrationally excited molecules is about $2600\,cm^{-1}$ lower in energy than that for ground vibrational state molecules. Using *ab initio* calculations of portions of the ground and excited PESs, the vibrational wavefunctions were calculated and the electronic excitation spectra simulated, using the overlap integral for the bound and dissociative vibrational wavefunctions on the two surfaces. Qualitative agreement of the calculation with the measurement suggests that at the energy of the fundamental vibration the O–H stretch is largely uncoupled from the rest of the molecule during the dissociation. The lack of coupling to the rest of the molecule is in line with the findings of the above-mentioned theoretical work[84] where coupling to the C–O bond was explored.

8.4.1.4   NH₂OH

One-photon dissociation and VMP of hydroxylamine, NH_2OH, pre-excited to the $4\nu_{OH}$ and $5\nu_{OH}$ vibrations were compared and discussed in terms of *ab initio* calculations.[86] Ground-state NH_2 and OH photoproducts were detected by LIF and electronically excited NH_2 *via* its fluorescence. The variety of dissociation channels accessible at total energies below $42\,000\,cm^{-1}$ that were applied in these experiments can best be presented graphically as in Figure 8.9. However, in all cases the $NH_2(\tilde{X}\,^2B_1) + OH(X\,^2\Pi)$ channel dominated, the other significant channel, $NH_2(\tilde{A}\,^2A_1) + OH(X\,^2\Pi)$, being only a minor one.

The *ab initio* calculations, which are in good agreement with the experimental observations, led to the conclusion that in both the one-photon and VMP experiments the S_1 singlet state is initially excited. Its character, however, changes drastically close to the FC region, where the lowest-excited A' and A'' states cross. The A' state correlates with $NH_2(\tilde{A}\,^2A_1) + OH(X\,^2\Pi)$ and the A'' state with $NH_2(\tilde{X}\,^2B_1) + OH(X\,^2\Pi)$. The most striking feature of the VMP experiments is the absence of any vector correlation, in particular the loss of rotational alignment and Λ-doublet preference. This is in contrast to the one-photon dissociation at 240 nm, where torsional excitation during the dissociation step leads to rotational alignment of the OH fragments and a preferential population of the $\Pi\,(A'')$ component of the Λ-doublet. Both are lost after isoenergetic two-photon excitation *via* O–H stretching overtones of NH_2OH. In addition, in the 240-nm dissociation most of the excess energy ($20\,550\,cm^{-1}$) is released into relative translation (53%) and NH_2 internal energy (40%, mostly

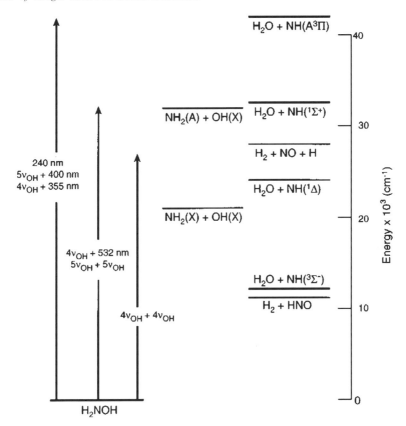

Figure 8.9 Dissociation channels accessible at total energies below $42\,000\,\text{cm}^{-1}$ in the VMP experiments on hydroxylamine. Reprinted with permission from Ref. 86. Copyright 1999, American Institute of Physics.

vibrational). OH carries little internal energy (7%), most of it in the form of rotational excitation. In the VMP experiments, higher internal excitation of the NH_2 fragments ($\sim 50\%$) is observed, at the expense of relative translation. At lower total excitation energies the relative translation takes up an increasing fraction of the total excess energy ($\geq 80\%$ at $5820\,\text{cm}^{-1}$ of excess energy).

8.4.1.5 *HO$_2$NO$_2$*

The interest in peroxynitric acid (PNA), HO_2NO_2, stems from its role in atmospheric chemistry.[88] PNA is formed in the atmosphere *via* a three-body recombination reaction involving the HO_2 and NO_2 radicals and, hence, provides a temporary reservoir for these species. Apart from its reaction with OH, photodissociation constitutes an important removal mechanism for atmospheric HO_2NO_2. Vibrationally excited PNA in the vicinity of its first OH-stretching overtone ($2\nu_1$) was dissociated *via* electronic excitation using 390-nm light.[88] The nascent energy distribution of the resulting OH fragments was probed using LIF at sub-Doppler resolution. It was found that for the minor $OH + NO_3$ channel, 55% of the $\sim 19\,600\,cm^{-1}$ available energy goes into relative translation of the two fragments and 45% into internal excitation. In addition, electronically excited NO_2 associated with the dominant $HO_2 + NO_2$ channel was also observed. The NO_2 fluorescence intensity was sufficiently intense to allow action spectra of both the $2v_1$ and $v_1 + 2v_3$ bands to be recorded. Comparing integrated intensities of these bands appearing in the state-selected action spectra with their known infrared absorption cross section enabled determination of the quantum yield for unimolecular dissociation for the predissociative $2v_1$ level. The quantum yield for unimolecular dissociation of the HO_2NO_2 $2v_1$ state was determined to be $\sim 30 \pm 5\%$ in the absence of collisions at 298 K.

8.4.1.6   CH₃OOH

Like HO_2NO_2, the interest in methyl hydroperoxide (MHP), CH_3OOH, stems from its role in atmospheric chemistry.[88a] It is readily photodissociated by near-UV light to produce OH and its traces in the atmosphere serve as a reservoir for OH and HO_2 radicals. In the VMP study of MHP the OH-stretching overtones and OH stretch–HOOC torsion (about the O–O bond) combination bands prepared in the $2v_{OH}$–$5v_{OH}$ regions were investigated using action spectroscopy.[88a] Results for the room-temperature spectra suggest that the coarse vibrational structures appearing in the spectra can be understood to a large extent by using a simple 2D vibration–torsion model involving the OH stretch and HOOC torsion. However, investigation of the jet-cooled action spectrum for the $2v_{OH}$ band along with results of *ab initio* calculations reveals that the dependence of the transition dipole moment on the HOOC torsion angle cannot be neglected when simulating intensities of the OH-stretching overtone bands. The calculations show that transitions between torsional levels of different symmetries contribute to a significant degree to the intensities of the corresponding OH-stretch overtone bands. These transitions are not allowed by a simple FC model but occur when the proper dependence of the dipole moment $\mu(r,\tau)$ on the torsional angle τ is accounted for.

8.4.2   Acetylene Homologues

VMP studies of three acetylene homologues, (normal) propyne ($H_3CC\equiv CH$),[17,94,95,97,98] propyne-d_3 ($D_3CC\equiv CH$)[36,89,90] and 1-butyne ($H_3CH_2CC\equiv CH$),[91–93,96,98] were carried out. Their presentation is hereafter divided into two sections: the first (8.4.2.1) concentrates on photodissociation dynamics aspects and the second (8.4.2.2) on applications to vibrational spectroscopy of which only the highlights are briefly reported. In these studies jet-cooled action spectra were compared to room-temperature PA spectra and to simulated spectra. The photodissociation was carried out at ~ 243.1 nm, the ensuing H(D) photofragments being monitored at this wavelength (243.135 and

243.069 nm for H and D $(2+1)$ REMPI through the $2s\,^2S \leftarrow 1s\,^2S$ two-photon transition).

8.4.2.1 Photodissociation Dynamics

Propyne and 1-butyne contain two chemically distinct types of C–H bond, acetylenic and alkylic (in 1-butyne there are two types of alkylic bonds) and it is of interest to find out which of these bonds is photolyzed at a given excitation energy and if it depends, despite fast IVR, on pre-excitation of different modes. For propyne the C–H bond dissociation energies of the acetylenic and methylic bonds are $\sim 45\,540$ and $\sim 31\,230\,cm^{-1}$, respectively.[102] It was therefore surprising that several studies in the 1990s concluded that photodissociation at 193.3 nm of vibrationless ground-state $H_3CC\equiv CD$,[103] $D_3CC\equiv CH$[104] and $H_3CC\equiv CH$[104–106] result in cleavage of the acetylenic C–H(D) bond rather than the much weaker methylic bond. Indeed, the results presented below of later photodissociation studies, both of vibrationally pre-excited propyne isotopologues[17,36,89,90,94,95,97,98] and of vibrationless ground-state species,[107–109,109a] *contradicted* the results of the previous studies. In sections *(a)* and *(b)* below we compare the results of the VMP studies of propyne-d_3 and propyne, respectively, to those of one-photon dissociation studies and in *(c)* we do the same for 1-butyne.

(a) $D_3CC\equiv CH$. VMP of $D_3CC\equiv CH$ (propyne-d_3)[36,89,90] pre-excited to the regions of the third $(4\nu_{CD})$ methyl overtone and to second $(3\nu_1)$ and third $(4\nu_1)$ acetylenic C–H overtone, with respective combined energies around 49 845, 50 810 and 53 880 cm^{-1} was investigated. These energies are slightly below and above the energy of 193.3-nm photons, 51 730 cm^{-1}, which was used for photodissociation of propyne isotopologues in the above-mentioned previous studies.[103–106] Detection of the H and D photofragments *via* Doppler and action spectroscopy[36,89,90] showed that the production of both fragments is, mainly, with low but similar translational energies and with propensities toward D. These features of the H and D photofragments in the VMP studies agree very well with those reported on photolysis at ~ 203–213 nm[107] and 193.3 nm[108] of propyne and propyne-d_3, but contradict the previous studies.[103–106] In particular, the statement of Ref. 103, that $H_3CC\equiv CD$ on 193.3-nm photodissociation yields only D atoms in spite of the fact that the C–H bond is much weaker than the C–D bond, seems to be in sharp contrast to the preference of D production in the nearly isoenergetic VMP of $D_3CC\equiv CH$ where the C–H stretch is pre-excited.[36,89,90]

In line with the conclusions of Ref. 107 and 108 (see more details in the next section), the translational energy disposal and the preference of D production in the VMP studies were suggested[90] to be the result of a fragmentation process preceded by IC to the ground-state PES, *isomerization to allene* (e.g. to $D_2C=C=CHD$) and finally unimolecular decay. For molecules pre-excited to the $3\nu_1$ and $4\nu_1$ acetylenic C–H stretch, an additional minor channel was observed[90] and considered to be a result of H-atom elimination from the acetylenic C–H bond, directly on the upper excited electronic state.

(b) $H_3CC{\equiv}CH$. VMP of $H_3CC{\equiv}CH$ (propyne) was studied for species initially excited to the regions of the first through the fourth overtone of the C–H methyl stretches[17,94,97,98] as well as to the second and third C–H acetylenic stretches.[95,98] The translational energy disposal, obtained *via* detection of the H photofragments applying Doppler spectroscopy, led to the conclusion that regardless of the pre-excited mode, the ensuing H atoms originate from cleavage of the methylic CH moiety. This suggests a similar dissociation mechanism to the above-mentioned mechanism for $D_3CC{\equiv}CH$.[36,89,90]

Some more details on the comparable studies of the one-photon dissociation of $H_3CC{\equiv}CH$,[107,108] relevant to the VMP work, are in place here. In these studies the photodissociation dynamics of jet-cooled $H_2C{=}C{=}CH_2$ (allene) and $H_3CC{\equiv}CH$ molecules following excitation at $\sim$203–213 nm[107] and 193.3 nm[108] was investigated by H-atom HRTOF PTS. The total kinetic-energy release spectra of the H atoms resulting from photolysis at a given wavelength in both molecules were found to be essentially identical. The form of these spectra could be reproduced reasonably well by a simple statistical model that assumed population of all possible vibrational states of the C_3H_3 partner, providing this partner was assumed to be the $H_2C{-}C{\equiv}CH$ (propargyl) radical. This could be rationalized if, for both molecules, IC to, and isomerization on, the S_0 state PES preceded fragmentation. The isomerization rate of the resulting highly vibrationally excited S_0 in both molecules was assumed to be faster than their unimolecular decay rate. It is worth noting that additional evidence contradicting the conclusion of the previous studies[103–106] of exclusive acetylenic bond fission was inferred from the formation of the $H_2C{-}C{\equiv}CH$ radical in the 193.3-nm photolysis of $H_3CC{\equiv}CH$.[109,109a]

The interpretation[95,97] of the energy disposal in VMP of $H_3CC{\equiv}CH$ is similar to that for the one-photon dissociation. The potential-energy levels of $H_3CC{\equiv}CH$ showing its isomerization to $H_2C{=}C{=}CH_2$ and the different vibrational pre-excitation schemes in the VMP experiments are depicted in Figure 8.10.

(c) $H_3CH_2CC{\equiv}CH$. The VMP of $H_3CH_2CC{\equiv}CH$ (1-butyne) was studied for the regions around the first through the third overtone of the C–H ethyl[91–93,98] and C–H acetylenic[96,98] stretches. The strengths of the two types of C–H bonds are $\sim$30 330 and $\sim$45 650 cm^{-1} for the ethylic[110] and acetylenic bond, respectively, the latter strength being estimated from that of the lower homologue, propyne.[102] In the photodissociation dynamics studies[92,93] all the ethylic hydrogens were considered to be similar due to lack of data, although the two C–H bonds in the methylene moiety might be somewhat weaker than that of the three in the methyl.

From the H-atom Doppler profile obtained following $\sim$243.1-nm photodissociation of 1-butyne pre-excited to the $2v_1$, $3v_1$ and $4v_1$ acetylenic stretches,

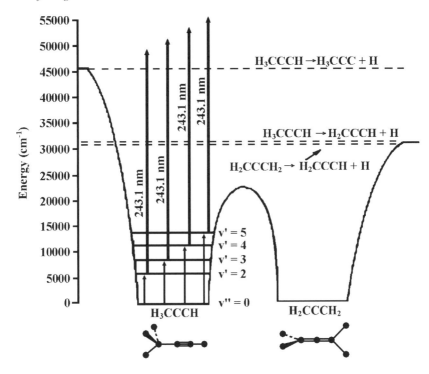

Figure 8.10 Schematics of the potential-energy levels of propyne (H₃CC≡CH) showing its isomerization to allene (H₂C=C=CH₂) and possible H-atom loss channels following 243.135-nm photodissociation of propyne initially excited to the first through the fourth overtone of the C–H methyl stretches (a similar isomerization mechanism is envisaged for VMP with initial excitation to the second and third C–H acetylenic stretches). The two lower dashed lines mark the bond dissociation energies for the preferential photodissociation channels and the upper dashed line for the acetylenic bond cleavage. Reprinted with permission from Ref. 94. Copyright 2005, American Institute of Physics.

the average translational energy $\langle Er \rangle$ of both fragments, $\sim 4200\,\mathrm{cm}^{-1}$, was calculated.[92,93] The respective combined energy (IR + UV) allocated to the parent is $\sim 47\,670$, $\sim 50\,810$ and $\sim 53\,880\,\mathrm{cm}^{-1}$. The respective available energy, E_{avl}, of the IR + UV photons over that required for C–H bond cleavage is $\sim 17\,340$, $\sim 20\,490$ and $\sim 23\,550\,\mathrm{cm}^{-1}$ for the ethylic and ~ 2160, ~ 5160 and $\sim 8230\,\mathrm{cm}^{-1}$ for the acetylenic bond. The E_{avl} values for the ethylic bond fissions are significantly higher than $\langle Er \rangle$, whereas for the acetylenic bonds they are lower, comparable and higher for $2v_1$, $3v_1$ and $4v_1$ pre-excitation, respectively. Since E_{avl} should *always* be higher than $\langle Er \rangle$, it is unlikely that significant release of hydrogen atoms occurs through their direct release from the acetylenic site. However, considering bond fission through the ethylic site indicates that the fraction of E_{avl} channeled into the photofragments' $\langle Er \rangle$, f_T, for photodissociation of 1-butyne initially excited to $2v_1$, $3v_1$ and $4v_1$ is 0.24, 0.20 and 0.18, respectively. These low values may indicate that the bond fission that presumably occurs through the ethylic bond does not take place directly on the upper PES but rather after IC to the ground electronic state. Moreover, if IC occurs another possibility could be raised, the subsequent isomerization of 1-butyne to 1,2-butadiene ($H_2C=C=CHCH_3$) or other forms of C_4H_6. We note that this resembles the isomerization of propyne to allene discussed above.

8.4.2.2   *Applications to Vibrational Spectroscopy*

In this section we briefly present some highlights of the extensive analysis of the vibrational spectroscopy of $H_3CC\equiv CH$ (propyne)[17,94,95,97,98] and $H_3CH_2CC\equiv CH$ (1-butyne) [91–93,96,98] based on their VMP studies. These molecules are appealing candidates for studying patterns of vibrational spectroscopy and vibrational energy flow since the two types of hydride oscillators, C–H acetylenic and alkylic stretches, may exhibit differing dynamics. The fundamental frequencies of the alkylic stretches are lower (around 2940 cm^{-1}) than the acetylenic C–H stretch (around 3335 cm^{-1}),[111–113] implying poor coupling between them. Coupling of the C–H acetylenic site to the rest of the molecule is also weak due to the 5:1 ratio between the frequencies of the C–H acetylenic stretching and bending. This ratio is high compared to the 2:1 stretch-deformation ratio, related to the Fermi resonance, in C–H oscillators of alkanes.

Vibrational transitions to hydride overtone oscillators, which are weak in direct excitation, were readily observed by action spectroscopy, even at the low densities of the molecular beam, due to favorable FC factors that lead to enhanced signals relative to those obtained from dissociation of vibrationless ground-state molecules. Comparison of action spectra of jet-cooled molecules to room-temperature photoacoustic (PA) spectra accompanied by simulations provided a means for distinguishing between the homogeneous and inhomo-geneous structure of vibrational transitions. The analysis of the overtone spectra of acetylenic C–H stretches and alkylic C–H stretches then provided a comparison between the two. Furthermore, analysis of the spectra of the alkylic C–H in terms of LM/NM and NM models brought additional insight into the role played by the intramolecular dynamics in IVR.

The main features of the vibrational spectra and intramolecular dynamics as obtained from the experiments and simulations can be summarized as follows.[98] The $3v_1$ and $3v_1 + v_3 + v_5$ bands in propyne, where $3v_1$ is the acetylenic stretch and v_3 and v_5 are the $C\equiv C$ and C–C stretching modes, include J resolved structures but largely unresolved K components, while the $4v_1$ acetylenic band, lying in the vicinity of the $3v_1 + v_3 + v_5$ combination band, loses its structure but still exhibited a recognizable band shape. This behavior implies that the $3v_1$ and $3v_1 + v_3 + v_5$ states are less strongly coupled to the bath states than the $4v_1$ state. In the larger 1-butyne molecule the rotational resolution was completely lost, but still the $2v_1$, $3v_1$ (and an adjacent newly observed state) and $4v_1$ acetylenic bands were characterized by well-defined shapes. Of much significance is also the observation that the linewidth of $4v_1$ in 1-butyne is smaller than that of the other two bands, implying a longer-lived $4v_1$ state. The very different linewidths in the spectra of the close-lying $3v_1 + v_3 + v_5$ and $4v_1$ states in propyne and the narrowing of the $4v_1$ band of 1-butyne point out that the energy decay out of the prepared state is not necessarily dependent on the density of states, but rather is affected by mode-selective coupling with the bath states or by the positions and couplings in the first several tiers of states.

The alkylic C–H stretches in propyne were characterized by well-defined band shapes, while in 1-butyne the shapes were almost completely lost. The diagonalization of simplified Hamiltonians enabled the eigenvalues and eigenvectors to be obtained[94,96,97] and the sets of parameters for each molecule to be retrieved by comparing the eigenvalues with the band origins *via* least-square analysis. It is the possibility of resolving many bands that allowed the parameters and consequently the Hamiltonian and the temporal evolution to be obtained. These findings imply that the decay of the population from the initially prepared pure LM states to the target states (combinations including bending) occurs on a subpicosecond timescale and is governed by Fermi resonances. Furthermore, in propyne it was possible to consider also the linewidths found from the simulations and the resulting corresponding exponential decays of the initially prepared states. These observations enabled the conclusion to be drawn out that the exponential decay almost does not affect the first overtone state, but rather the fourth one. This means that the excitation of methyl C–H stretch is rapidly lost *via* strong interactions of the former with deformations and couplings of the latter with bath states. At longer times the first overtone state also couples to bath states and its population decays.

8.4.3 Ethene Isotopologues

The simplest olefin, ethene (C_2H_4), serves as a paradigm for understanding the dynamics of excited states in molecules containing a carbon–carbon double bond. The VMP studies of the ethene isotopologues listed below[99–101] were aimed at providing new insights into their interesting photophysics and spectroscopy. VMP of jet-cooled species was compared to room-temperature PA spectra and to simulated spectra. Following photodissociation at ~ 243.1 nm, the yield of the ensuing H(D) photofragments was monitored *via* $(2+1)$ REMPI at the same wavelength. In (normal) ethene ($H_2C=CH_2$)[99,100] the first through fourth C–H stretching overtone regions, in 1,2-*trans*-d_2-ethene (*trans*-HDC=CDH)[99,101] the first and the fourth and in 1,1-d_2-ethene ($H_2C=CD_2$)[99] the fourth C–H stretching overtone region were studied. Before presenting more details, a couple of introductory remarks on one-photon dissociation studies of ethene and on its vibrational level structure are required.

Most relevant to the VMP studies are the measurements and computations of the branching ratios of the atomic photofragments in one-photon dissociation at comparable energies. In 193-nm photodissociation the H/D ratios were determined to be 2.2 ± 0.2 for $H_2C=CD_2$ and 1.2 ± 0.1 for *trans*-HDC=CDH.[114] It was suggested that IC precedes dissociation and the RRKM model was used to explain the observed ratios. These ratios differ from the computed values of 1.34 and 1.38 in Ref. 115 and 1.5 and 1.3 in 116. In particular, in Ref. 115 it was concluded that the atomic elimination is non-statistical, contrasting with the above observation.[114]

Another relevant issue is the vibrational level structure in the electronic ground state of ethene. Despite extensive spectroscopic work, the vibrational level structure has not been yet completely worked out. Although ethene is a relatively simple molecule, its spectra are considered to be one of the most difficult to explain. It is obvious that because of the size of the molecule, anharmonic resonances and Coriolis-type interactions figure extensively and complicate the spectral pattern. Structural resonances between the four C–H stretches are expected to occur,[117] in addition to the perturbation of the v_{11} mode (b_{1u} C–H stretch) by a lower-order anharmonic resonance with $v_2 + v_{12}$, corresponding to the C=C stretching and C–H in-plane bending, respectively. The pattern of couplings in ethene and its derivatives is quite different from that in other classes of small organic molecules, *e.g.*, methane and acetylene and their derivatives and therefore raises much interest.

In the following sections we briefly summarize the results of the VMP studies of the fourth C–H stretching overtone region in the above-mention three ethene isotopologues[99] (Section 8.4.3.1), the first through fourth C–H stretching overtone regions in $H_2C=CH_2$[100] (Section 8.4.3.2) and the first stretch overtone region in *trans*-HDC=CDH[101] (Section 8.4.3.3). Each section highlights some specific features in the VMP of ethene. However, following the detailed VMP-based analysis of the vibrational spectroscopy and dynamics of *trans*-HDC=CDH[101] we present the results of this analysis in more detail.

8.4.3.1   The Fourth C–H Stretching Overtone Region in $H_2C=CH_2$, *Trans*-HDC=CDH and $H_2C=CD_2$

The action spectra of the three isotopologues are partially resolved, due to the inhomogeneous structure reduction, and were found to be key players in identifying the characteristics of the spectral features.[99] The simultaneous simulations of the PA and action spectra allowed retrieval of the band types and origins, molecular constants and homogeneous linewidths and thus intramolecular dynamical information to be decoded. This was achieved by considering the positions and the linewidths of the involved bands, enabling estimation of the timescales for energy redistribution to the doorway and to the bath states. The limits for the redistribution times from the initially excited C–H stretch to possible doorway states were estimated to be <0.5 ps for $H_2C=CH_2$, <0.4 ps for $H_2C=CD_2$ and <2.4 ps for *trans*-HDC=CDH, respectively. The longer time for the last molecule is, apparently, due to its relatively isolated C–H bonds.

The Doppler profiles indicated that the translational energy disposal in the H and D photofragments is low. This was attributed to the occurrence of pho-tofragmentation *via* IC to the (hot) ground-state PES followed by unimolecular decay to H(D) and corresponding vinyl photofragments. Of particular note are the observed propensities toward H photofragments in $H_2C=CD_2$ and *trans*-HDC=CDH, with H/D branching ratios of 1.8 ± 0.3 and 1.6 ± 0.2, respectively, which reflect the trend found in the corresponding calculated ratios of 1.5 and 1.3 in Ref. 116. The preferential C–H bond cleavage seems to be due to a dynamical effect and not to the slightly different photodissociation wavelengths used for monitoring of the H and D photofragments (243.135 and 243.069 nm, respectively, which are about $11\,\mathrm{cm}^{-1}$ apart, out of the combined energy of $\sim 55\,250\,\mathrm{cm}^{-1}$). Since energy redistribution occurs in picoseconds, it is likely that the memory of initial vibrational excitation is lost prior to the UV excitation, yet it might be that the FC overlap with the upper PES, or the transition moment to the C–H dissociation channel is favorable relative to C–D and the light H atom escapes more effectively. It was also suggested[114] that the isotope effect for H and D elimination from $H_2C=CD_2$ could be more prominent than from *trans*-HDC=CDH due to the different nature of their activated complexes.

8.4.3.2 Vibrational Patterns of the First Through Fourth C–H Stretching Overtone Regions in $H_2C=CH_2$

The study of the first through fourth C–H stretch overtone regions of $H_2C=CH_2$ was aimed at the investigation of its vibrational patterns.[100] The room-temperature PA and jet-cooled action spectra were characterized by A- and B-type bands only, which correspond to states of B_{1u} and B_{2u} symmetry, respectively. This could be realized by considering that ethene is an asymmetric rotor of planar geometry in the ground state, belonging to the D_{2h} point group. Provided that the rotational selection rules and thus the band types observed in the vibrational spectrum are set by the transition dipole moment of the IR active C–H stretch bright states, it is reasonable to find components of these types only.

The measured PA and action spectra were simulated taking advantage of the better resolution of the latter. These simulations allowed retrieval of the multiple band structure of each manifold and the characteristic linewidths, which were found to increase with increasing quantum number of the C–H stretch. Analysis of these multiband-structured spectra in terms of a simplified joint LM/NM model and an equivalent NM model, which accounted for the Fermi resonances with the overtones of the CH_2 scissoring ($2v_3$, $2v_{12}$) and combination of the scissoring (v_3) with the C=C stretch (v_2), enabled eigenvalues and eigenvectors to be obtained.

By comparison and fitting of the eigenvalues to the experimental band positions and by consideration of the LM character and the observed intensity of the bands, the assignment of the observed bands could be attained. Furthermore, some mechanistic insight regarding energy flow out of the initially excited C–H stretches was also obtained. The B_{1u} and B_{2u} symmetry bright states, respectively, behave differently through the $V = 2$–4 manifolds (first through third C–H stretch overtone), showing coupling to doorway states for the former, but not for the latter. The initial energy flow out of the B_{1u} states is to the doorway states *via* Fermi resonances and occurs on subpicosecond timescales. Of particular importance is the fact that the density of states for the $V = 2$–4 manifolds was not sufficient to be pronounced by the measured line-widths under the given resolution. On the other hand, for $V = 5$ the higher density of states causes extensive homogeneous broadening, which reflects the energy decay out of the C–H stretches of the fourth overtone to the bath states.

8.4.3.3   The First Stretch Overtone Region in Trans-HDC=CDH: Example of VMP-Based Detailed Analysis of Vibrational Spectroscopy and Dynamics

In comparison to $H_2C=CH_2$, *trans*-HDC=CDH has been previously only scantly studied. We therefore present here a VMP-based analysis of the vibrational spectroscopy and dynamics of the latter in more detail than that of the other ethene isotopologues.

Vibrational pre-excitation in the first C–H stretch overtone region of *trans*-HDC=CDH was followed by excitation to the first electronic state, dissociation and eventual H and D photofragment release by C–H and C–D bond fission, respectively. Monitoring of the PA and H and D action spectra and simulations of the spectra enabled modeling of the observed multiband structure and the principal resonances and the IVR mechanisms to be revealed.

To calculate the vibrational NMs and to help in the vibrational analysis of the monitored spectra, *ab initio* and DFT calculations were used. Harmonic frequencies, obtained by *ab initio* methods, were combined with DFT anhar-monic corrections in a relatively small basis set.[101] The modes, their assign-ments, symmetries and nuclear displacements, are presented in Figure 8.11. Also indicated are the calculated frequencies and for comparison the corre-sponding experimental values[118] that agree very well with the former. All NMs were successfully assigned to one of four type symmetries by considering the displayed motions.[101] The molecule belongs to the C_{2h} symmetry and, as is apparent from Figure 8.11, it has two A_u and four B_u IR active fundamentals giving rise to C-type bands and to hybrid $A{:}B$ bands, respectively.

The PA and H and D action spectra in the energetic window of the first C–H stretch overtone region showed an intense dominant band around $5990\,\mathrm{cm}^{-1}$ and five weaker bands in the 5900–$6220\,\mathrm{cm}^{-1}$ region. Whereas the action spectra allowed resolution of the band at $\sim 5990\,\mathrm{cm}^{-1}$, the PA spectrum enabled observation of two bands at $\sim 6219\,\mathrm{cm}^{-1}$, which could not be traced by the action spectra due to low FC overlap with the upper vibronic states. From the fundamental frequencies of the NMs and their symmetries, presented in Figure 8.11, it is obvious that the ~ 5990-cm^{-1} dominant band must be related to the $v_1 + v_9$ combination of the first C–H stretch overtone.

To interpret and assign the spectral features, the $v_1 + v_9$ state was considered as the zero-order bright state (ZOBS) and the other five states as zero-order dark

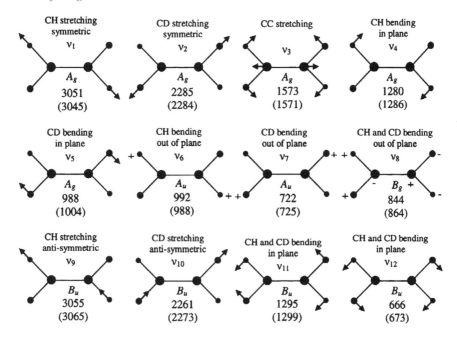

Figure 8.11 The twelve NMs of vibration of *trans*-HDC=CDH (1,2-*trans*-d_2-ethene) as obtained from calculated harmonic frequencies with anharmonic corrections.[101] Below each calculated frequency the experimental frequency[118] is given in parentheses. The numbering of the modes is according to their energies and symmetries. Reprinted with permission from Ref. 101. Copyright 2008, American Institute of Physics.

states (ZODS) coupled to it. Accordingly, a Hamiltonian was constructed, consisting of diagonal zero-order energy terms, corresponding to the $v_1 + v_9$ ZOBS and to the five ZODS and off-diagonal interaction matrix elements between them, while the interactions between the ZODS themselves were assumed to be negligible. The unperturbed energies and interaction elements were retrieved by numerical solution, through an iterative matrix diagonalization and nonlinear least squares fit routine, which fitted the eigenvalues to the band origins, and the squared eigenvector coefficients of the ZOBS to the normalized band intensities. Explicitly, 11 parameters (6 zero-order and 5 off-diagonal matrix elements) were fitted to the 6 measured energies and 5 relative intensities. The resulting values, corresponding to the energies of the states and the matrix elements are given, on the energy axis and in the vicinity of the arrows of Figure 8.12, respectively. Evidently, the matrix elements for the $v_1 + v_9$ interaction with four of the ZODS are tens of cm^{-1}, while with the fifth one only ~ 3.4 cm^{-1}.

Taking into account the experimental fundamental frequencies (Figure 8.11) and assuming that the ZODS must have the same symmetry as the $v_1 + v_9$ ZOBS and that only low-order resonances might be involved in their interaction, the energies of coupling candidate states, in the 5800–6300 cm^{-1} range, were calculated. Comparing the calculated frequencies to those of the ZODS,

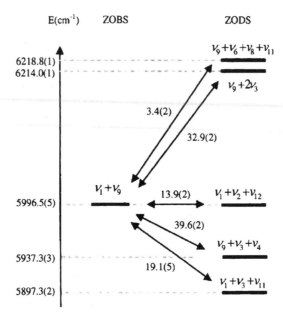

Figure 8.12 Schematic representation of the deperturbed states of B_u symmetry in the first C–H stretch overtone region of *trans*-HDC=CDH (1,2-*trans*-d_2-ethene). The horizontal solid lines indicate the zero-order state position and the arrows show the interactions. Also given, on the energy axis, are the state positions with their accuracies in the less significant digit (in parentheses), their assignment (above the solid lines) and the absolute values of the matrix elements in cm^{-1} (near the arrows). Reprinted with permission from Ref. 101. Copyright 2008, American Institute of Physics.

Figure 8.12, the assignment of the ZODS was deduced and is presented in the figure.

Based on the measured spectra and the derived Hamiltonian, the temporal behavior of the states was computed.[101] It was found that the population of the $v_1 + v_9$ state decays very fast, < 0.2 ps, and then continues to oscillate. The other states acquire population, each in its turn, and oscillate with characteristic periods. The $v_9 + 2v_3$ and $v_9 + v_6 + v_8 + v_{11}$ states exhibit oscillations with longer periods (~ 30 ps) in addition to the very fast oscillations, showing that when the former is populated the latter has no population and *vice versa*. We note that the presented temporal behavior corresponds to the picture obtained due to the strong interactions only. It did not take into account the weak interactions with the bath states of B_u symmetry that are very sparse in this energetic window, < 1 state/cm^{-1}. This is since *trans*-HDC=CDH, like $H_2C=CH_2$, is characterized by a unique behavior where the density of states in the region of the lower overtones is relatively low. This, in turn, is due to the rather high energy of the fundamentals, where the lowest, 673 cm^{-1} (see Figure 8.11), is related to the C–H and C–D in-plane bending.

8.4.4 UF$_6$

To conclude our presentation of VMP of "larger than tetratomic" molecules we turn to uranium hexafluoride, UF$_6$. The dissociation of this molecule applying IR multiphoton excitation was of much interest in the 1970s and 1980s in the context of uranium isotope separation, but application of VMP for this purpose was considered as well. We quote here two works that are related to VMP of UF$_6$. In the earlier one[119] an investigation was made of the absorption of UV in UF$_6$ following IR pre-excitation of the combination vibration $v_2 + v_3$ at 1157 cm^{-1}. A large increase in the absorption coefficient in the 250–550 nm range and a considerable redshift of the photoabsorption edge were observed. The ratio between the UV absorption coefficient of the excited and unexcited molecules exceeded 50 at the wavelengths of 340 and 420–550 nm. It was suggested that the considerable redshift may indicate multi-quantum vibrational pre-excitation.[119] In the later work[120] the absorption spectrum of UF$_6$ between the wavelengths 330 and 450 nm in the temperature range 295 to 500 K was examined in order to obtain some idea of the effect of vibrational excitation on the absorption cross sections in this region. The enhancement of absorption with temperature was consistent with the predicted change in the Boltzmann population distribution in the vibrational levels of the ground state. However, the upper bound of the photodissociation cross section was found to be small and the authors concluded that UV photodissociation does not appear to be a promising method of discriminating the vibrationally excited UF$_6$ molecules.

References

1. R. V. Ambartzumian and V. S. Letokhov, *Chem. Phys. Lett.*, 1972, **13**, 446.
1(a). J. S. Lim, Y. S. Lee and S. K. Kim, *Angew. Chem. Int. Ed.*, 2008, **47**, 1853.
1(b). G. Herzberg, *Molecular Spectra and Molecular Structure III, Electronic Spectra and Electronic Structure of Polyatomic Molecules*, Van Nostrand Reinhold, New York, 1966.
2. J. V. Michael and W. A. Noyes, *J. Am. Chem. Soc.*, 1963, **85**, 1228.
3. E. Kassab, J. T. Gleghorn and E. M. Evleth, *J. Am. Chem. Soc.*, 1983, **105**, 1746.
4. G. C. G. Waschewsky, D. C. Kitchen, P. W. Browning and L. J. Butler, *J. Phys. Chem.*, 1995, **99**, 2635.
5. K. M. Dunn and K. Morokuma, *J. Phys. Chem.*, 1996, **100**, 123.
6. C. L. Reed, M. Kono and M. N R. Ashfold, *J. Chem. Soc. Faraday Trans.*, 1996, **92**, 4897.
7. M. N. R. Ashfold, R. N. Dixon, M. Kono, D. H. Mordaunt and C. L. Reed, *Philos. Trans. R. Soc. Lond. A*, 1997, **355**, 1659.
8. S. J. Baek, K.-W. Choi, Y. S. Choi and S. K. Kim, *J. Chem. Phys.*, 2002, **117**, 10057.

9. S. J. Baek, K.-W. Choi, Y. S. Choi and S. K. Kim, *J. Chem. Phys.*, 2003, **118**, 11026.

10. M. H. Park, K.-W. Choi, S. Choi, S. K. Kim and Y. S. Choi, *J. Chem. Phys.*, 2006, **125**, 084311.

10(a). D.-S. Ahn, J. Lee, J.-M. Choi, K.-S. Lee, S. J. Baek, K. Lee, K.-K. Baeck and S. K. Kim, *J. Chem. Phys.*, 2008, **128**, 224305.

11. T. Shimanouchi, *Tables of Molecular Vibrational Frequencies Consolidated Volume I*, National Bureau of Standards, Washington D.C., 1972.

12. M. Kreglewski, *J. Mol. Spectrosc.*, 1978, **72**, 1.

13. M. Kreglewski, *J. Mol. Spectrosc.*, 1989, **133**, 10.

14. D. S. Perry, G. A Bethardy and X. L. Wang, *Ber. Bunsen. Ges. Phys. Chem.*, 1995, **99**, 530.

15. A. Golan, S. Rosenwaks and I. Bar, *J. Chem. Phys.*, 2006, **125**, 151103.

16. A. Golan, S. Rosenwaks and I. Bar, *Isr. J. Chem.*, 2007, **47**, 11.

17. A. Golan, A. Portnov, S. Rosenwaks and I. Bar, *Phys. Scr.*, 2007, **76**, C79.

18. R. Marom, U. Zecharia, S. Rosenwaks and I. Bar, *Chem. Phys. Lett.*, 2007, **440**, 194.

19. R. Marom, U. Zecharia, S. Rosenwaks and I. Bar, *Mol. Phys.*, 2008, **106**, 213.

20. R. Marom, U. Zecharia, S. Rosenwaks and I. Bar, *J. Chem. Phys*, 2008, **128**, 154319.

20(a). C. Levi, J. M. L. Martin and I. Bar, *J. Comput. Chem.*, 2008, **29**, 1268.

20(b). C. Levi, G. J. Halász, Á. Vibók, I. Bar, Y. Zeiri, R. Kosloff and M. Baer, *J. Chem. Phys.*, 2008, **128**, 244302.

21. F. S. Rowland, *Environ. Sci. Technol.*, 1991, **25**, 622.

22. M. McFarlan and J. Kaye, *Photochem. Photobiol.*, 1992, **55**, 911.

23. X. Yang, P. Felder and J. R. Huber, *Chem. Phys.*, 1994, **189**, 127.

24. A. Melchior, P. Knupfer, I. Bar, S. Rosenwaks, T. Laurent, H.-R. Volpp and J. Wolfrum, *J. Phys. Chem.*, 1996, **100**, 13375.

25. A. Melchior, I. Bar and S. Rosenwaks, *J. Chem. Phys.*, 1997, **107**, 8476.

26. A. Melchior, H. M. Lambert, P. J. Dagdigian, I. Bar and S. Rosenwaks, *Isr. J. Chem.*, 1997, **37**, 455.

27. R. A. Brownsword, M. Hillenkamp, T. Laurent, R. K. Vatsa, H.-R. Volpp and J. Wolfrum, *J. Phys. Chem. A*, 1997, **101**, 995.

28. R. A. Brownsword, M. Hillenkamp, T. Laurent, H.-R. Volpp, J. Wolfrum, R. K. Vatsa and H.-S. Yoo, *J. Chem. Phys.*, 1997, **107**, 779.

29. R. A. Brownsword, P. Schmiechen, H.-R. Volpp, H. P. Upadhyaya, Y. J. Jung and K.-H. Jung, *J. Chem. Phys.*, 1999, **110**, 11823.

30. L.-H. Lai, Y.-T. Hsu and K. Liu, *Chem. Phys. Lett.*, 1999, **307**, 385.

31. A. Melchior, X. Chen, I. Bar and S. Rosenwaks, *Chem. Phys. Lett.*, 1999, **315**, 421.

32. A. Melchior, X. Chen, I. Bar and S. Rosenwaks, *J. Chem. Phys.*, 2000, **112**, 10787.

33. A. Melchior, X. Chen, I. Bar and S. Rosenwaks, *J. Phys. Chem. A*, 2000, **104**, 7927.

34. X. Chen, R. Marom, S. Rosenwaks, I. Bar, T. Einfeld, C. Maul and K-H. Gericke, *J. Chem. Phys.*, 2001, **114**, 9033.
35. T. Einfeld, C. Maul, K.-H. Gericke, R. Marom, S. Rosenwaks and I. Bar, *J. Chem. Phys.*, 2001, **115**, 6418.
36. I. Bar and S. Rosenwaks, *Int. Rev. Phys. Chem.*, 2001, **20**, 711.
37. L. Li, G. Dorfman, A. Melchior, S. Rosenwaks and I. Bar, *J. Chem. Phys.*, 2002, **116**, 1869.
38. G. Dorfman, A. Melchior, S. Rosenwaks and I. Bar, *J. Phys. Chem. A*, 2002, **106**, 8285.
39. H. Okabe, *Photodissociation of Small Molecules*, John Wiley & Sons, New York, 1978.
40. H.-R. Dübal and M. Quack, *Mol. Phys.*, 1984, **53**, 257.
41. K. Tonokura, Y. Matsumi, M. Kawasaki, S. Tasaki and R. Bersohn, *J. Chem. Phys.*, 1990, **93**, 7981.
42. Y. Matsumi, K. Tonokura, M. Kawasaki, K. Tsuji and K. Obi, *J. Chem. Phys.*, 1993, **98**, 8330.
43. R. Liyange, Y. Yang, S. Hashimoto, R. J. Gordon and R. W. Field, *J. Chem. Phys.*, 1995, **103**, 6811.
44. P. M. Regan, S. R. Langford, D. Ascenzi, P. A. Cook, A. J. Orr-Ewing and M. N. R. Ashfold, *Phys. Chem. Chem. Phys.*, 1999, **1**, 3247.
45. R. N. Zare, *Mol. Photochem.*, 1972, **4**, 1.
46. G. E. Hall and P. L. Houston, *Annu. Rev. Phys. Chem.*, 1989, **40**, 375.
47. R. J. Gordon and G. E. Hall, *Adv. Chem. Phys.*, 1996, **96**, 1.
48. C. Maul and K.-H. Gericke, *Int. Rev. Phys. Chem.*, 1997, **16**, 1.
49. C. Maul and K.-H. Gericke, *J. Phys. Chem. A*, 2000, **104**, 2531.
50. J. F. Ying and K. T. Leung, *J. Chem. Phys.*, 1996, **105**, 2188.
51. M. Shapiro, *Chem. Phys. Lett.*, 1981, **81**, 521.
52. M. Shapiro and R. Bersohn, *Ann. Rev. Phys. Chem.*, 1982, **33**, 409.
53. M. D. Barry and P. A. Gorry, *Mol. Phys.*, 1984, **52**, 461.
54. M. Shapiro, *J. Phys. Chem.*, 1986, **90**, 3644.
55. H. Guo, K. Q. Lao, G. C. Schatz and A. D. Hammerich, *J. Chem. Phys.*, 1991, **94**, 6562.
56. H. Guo, *J. Chem. Phys.*, 1992, **96**, 2731.
57. C. Rist and M. H. Alexander, *J. Chem. Phys.*, 1993, **98**, 6196.
57(a). M. Shapiro, in *Isotope Effects in Gas-Phase Chemistry*, ed. J.A. Kaye, ACS Symposium Series 502, ACS, Washington D.C., 1992, p. 264.
58. A. T. J. B. Eppink and D. H. Parker, *J. Chem. Phys.*, 1999, **110**, 832.
59. I. Thanopulos and M. Shapiro, *J. Chem. Phys.*, 2006, **125**, 133314.
59(a). I. Thanopulos and M. Shapiro, *J. Phys. B*, 2008, **41**, 074010.
60. A. B. Alekseyev, H. P. Liebermann, R. J. Buenker and S. N. Yurchenko, *J. Chem. Phys.*, 2007, **126**, 234102.
61. A. B. Alekseyev, H. P. Liebermann and R. J. Buenker, *J. Chem. Phys.*, 2007, **126**, 234103.
62. R. S. Mulliken, *J. Chem. Phys.*, 1940, **8**, 382.
63. H. Pummer, J. Eggleston, W. K. Bischel and C. K. Rhodes, *Appl. Phys. Lett.*, 1978, **32**, 427.

64. P. F. Zittel and D. D. Little, *Chem. Phys.*, 1981, **63**, 227.

65. M. D. Person, P. W. Kash and L. J. Butler, *J. Chem. Phys.*, 1991, **94**, 2557.

66. M. Hippler and M. Quack, *Chem. Phys. Lett.*, 1994, **231**, 75.

67. M. Hippler and M. Quack, *Ber. Bunsenges. Phys. Chem.*, 1995, **99**, 417.

68. M. Hippler and M. Quack, *J. Chem. Phys.*, 1996, **104**, 7426.

69. H. M. Lambert and P. J. Dagdigian, *Chem. Phys. Lett.*, 1997, **275**, 499.

70. H. M. Lambert and P. J. Dagdigian, *J. Chem. Phys.*, 1998, **109**, 7810.

71. C. Tao and P. J. Dagdigian, *Chem. Phys. Lett.*, 2001, **350**, 63.

72. R. Marom, A. Golan, S. Rosenwaks and I. Bar, *Chem. Phys. Lett.*, 2003, **378**, 305.

73. R. Marom, A. Golan, S. Rosenwaks and I. Bar, *J. Phys. Chem. A*, 2004, **108**, 8089.

74. H.-Y. Xiao, Y.-J. Liu, J.-G. Yu and W.-H. Fang, *Chem. Phys. Lett.*, 2007, **436**, 75.

75. J. E. Baggott, D. A. Newnham, J. L. Duncan, D. C. McKean and A. Brown, *Mol. Phys.*, 1990, **70**, 715.

76. M. M. Law and J. L. Duncan, *Mol. Phys.*, 1998, **93**, 809.

77. Z. Lan, W. Domcke, V. Vallet, A. L. Sobolewski and S. Mahapatra, *J. Chem. Phys.*, 2005, **122**, 224315.

78. M. G. D. Nix, A. L. Devine, B. Cronin, R. N. Dixon and M. N. R. Ashfold, *J. Chem. Phys.*, 2006, **125**, 133318.

79. M. L. Hause, Y. H. Yoon, A. S. Case and F. F. Crim, *J. Chem. Phys.*, 2008, **128**, 104307.

80. A. Sinha, R. L. Vander Wal, L. J. Butler and F. F. Crim, *J. Phys. Chem.*, 1987, **91**, 4645.

81. M. D. Likar, J. E. Baggott, A. Sinha, T. M. Ticich, R. L. Vander Wal and F. F. Crim, *J. Chem. Soc., Faraday Trans. 2*, 1988, **84**, 1483.

82. M. D. Likar, J. E. Baggott and F. F. Crim, *J. Chem. Phys.*, 1989, **90**, 6266.

83. A. Sinha, R. L. Vander Wal and F. F. Crim, *J. Chem. Phys.*, 1989, **91**, 2929.

84. C. C. Marston, K. Weide, R. Schinke and H. U. Suter, *J. Chem. Phys.*, 1993, **98**, 4718.

85. D. Bingemann, M. P. Gorman, A. M. King and F. F. Crim, *J. Chem. Phys.*, 1997, **107**, 661.

86. D. Luckhaus, J. L. Scott and F. F. Crim, *J. Chem. Phys.*, 1999, **110**, 1533.

87. J. M. Hutchison, R. J. Holiday, A. Bach, S. Hsieh and F. F. Crim, *J. Phys. Chem. A*, 2004, **108**, 8115.

88. J. Matthews, R. Sharma and A. Sinha, *J. Phys. Chem. A*, 2004, **108**, 8134.

88(a). J. Matthews, M. Martinez-Avilés, J. S. Francisco and A. Sinha, *J. Chem. Phys.*, 2008, **129**, 074316.

89. X. Chen, Y. Ganot, I. Bar and S. Rosenwaks, *J. Chem. Phys.*, 2000, **113**, 5134.

90. Y. Ganot, S. Rosenwaks and I. Bar, *J. Chem. Phys.*, 2004, **120**, 8600.

91. A. Portnov, S. Rosenwaks and I. Bar, *Chem. Phys. Lett.*, 2004, **392**, 140.

92. A. Portnov, S. Rosenwaks and I. Bar, *J. Chem. Phys.*, 2004, **121**, 5860.
93. A. Portnov, Y. Ganot, S. Rosenwaks and I. Bar, *J. Mol. Struct.*, 2005, **744–747**, 107.
94. A. Portnov, E. Bespechansky, Y. Ganot, S. Rosenwaks and I. Bar, *J. Chem. Phys.*, 2005, **122**, 224316.
95. Y. Ganot, S. Rosenwaks and I. Bar, *J. Chem. Phys.*, 2005, **122**, 244318.
96. A. Portnov, E. Bespechansky, S. Rosenwaks and I. Bar, *J. Chem. Phys.*, 2005, **123**, 084316.
97. A. Portnov, L. Blockstein and I. Bar, *J. Chem. Phys.*, 2006, **124**, 164301.
98. A. Portnov, Y. Ganot, E. Bespechansky, S. Rosenwaks and I. Bar, *Vib. Spectrosc.*, 2006, **42**, 147.
99. E. Bespechansky, A. Portnov, A. Zwielly, S. Rosenwaks and I. Bar, *J. Chem. Phys.*, 2006, **125**, 133301.
100. A. Portnov, E. Bespechansky and I. Bar, *J. Phys. Chem. A*, 2007, **111**, 10646.
101. A. Zwielly, A. Portnov, C. Levi, S. Rosenwaks and I. Bar, *J. Chem. Phys.*, 2008, **128**, 114305.
102. M. S. Robinson, M. L. Polak, V. M. Bierbaum, C. H. DePuy and W. C. Lineberger, *J. Am. Chem. Soc.*, 1995, **117**, 6766.
103. S. Satyapal and R. Bersohn, *J. Phys. Chem.*, 1991, **95**, 8004.
104. K. Seki and H. Okabe, *J. Phys. Chem.*, 1992, **96**, 3345.
105. C.-K. Ni, J. D. Huang, Y. T. Chen, A. H. Kung and W. M. Jackson, *J. Chem. Phys.*, 1999, **110**, 3320.
106. W. Sun, K. Yokoyama, J. C. Robinson, A. G. Suits and D. M. Neumark, *J. Chem. Phys.*, 1999, **110**, 4363.
107. R. H. Qadiri, E. J. Feltham, E. E. H. Cottrill, N. Taniguchi and M. N. R. Ashfold, *J. Chem. Phys.*, 2002, **116**, 906.
108. R. H. Qadiri, E. J. Feltham, N. H. Nahler, R. P. García and M. N. R. Ashfold, *J. Chem. Phys.*, 2003, **119**, 12842.
109. J. D. DeSain and C. A. Taatjes, *J. Phys. Chem. A*, 2003, **107**, 4843.
109(a). J. C. Robinson, N. E. Sveum, S. J. Goncher and D. M. Neumark, *Mol. Phys.*, 2005, **103**, 1765.
110. Y. Hidaka, T. Higashihara, T. Oki and H. Kawano, *Int. J. Chem. Kinetics*, 1995, **27**, 321.
111. G. K. Speirs and J. L. Duncan, *J. Mol. Spectrosc.*, 1974, **51**, 277.
112. J. Saussey, J. Lamotte and J. C. Lavally, *Spectrochim. Acta A*, 1976, **32**, 763.
113. L. C. Baylor, E. Weitz and P. Hofmann, *J. Chem. Phys.*, 1989, **90**, 615.
114. S. Satyapal, G. W. Johnston, R. Bersohn and I. Oref, *J. Chem. Phys.*, 1990, **93**, 6398.
115. A. H. H. Chang, D. W. Hwang, X. M. Yang, A. M. Mebel, S. H. Lin and Y. T. Lee, *J. Chem. Phys.*, 1999, **110**, 10810.
116. A. P. Gallego, E. Martinez-Nunez and S. A. Vazquez, *Chem. Phys. Lett.*, 2002, **353**, 418.
117. M. Herman, J. Lievin, J. Vander Auwera and A. Campargue, *Adv. Phys. Chem.*, 1999, **108**, 1.

118. J. L. Duncan, D. C. McKean and P. D. Mallinson, *J. Mol. Spectrosc.*, 1973, **45**, 221.
119. R. V. Ambartzumian, I. G. Zubarev, A. A. Iogansen and A. V. Kotov, *Sov. J. Quantum Electron*, 1978, **8**, 910.
120. J. A. Horsley, P. Rabinowitz, A. Stein, D. M. Cox, R. O. Brickman and A. Kaldor, *IEEE J. Quantum Electron*, 1980, **QE-16**, 412.

Concluding Remarks

Throughout this monograph we have presented the methodology of VMP *via* specific examples of studies of isolated, neutral molecules in the gas phase by applying laser pulses in the ns timescale. We have learned that controlling the photodissociation products ensuing from an electronically excited state is achievable *via* pre-excitation of specific vibrational states in the initial electronic state. However, this mode selectivity has been found to work only for molecules not larger than tetratomic, with only one exception. The main reason for this limit is IVR. One could suggest that working in the sub-ns timescale may, at least in some cases, alleviate the problem of IVR, but there is a tradeoff between the pulse length and its spectral width. In large molecules, where IVR is fast, the spectral width of short pulses falls short of exciting specific vibrational states. In other words, when the size and complexity of the molecule increases, the IVR lifetime decreases and the spectral congestion increases. Still, it is surprising that thus far very few attempts of sub-ns VMP studies of medium-size molecules (around four to six atoms, say) have been reported.

We have also learned that VMP is an effective tool in molecular spectroscopy and molecular dynamics studies. It is effective, in particular, for determination of IVR lifetimes and for studying the vibrational spectroscopy of states that are difficult to study with other methods. The above-mentioned limit of the size of the molecule is irrelevant here. For observing the mode selectivity in VMP the vibrational excitation has to survive IVR in order to retain the selectivity since the subsequent electronic excitation has to be from the excited vibrational state. In contrast, monitoring vibrational molecular dynamics relies only on the efficacy of the excitation of the specific rovibrational state. When IVR is fast and rovibrational distribution reaches equilibrium, the subsequent electronic excitation will still reflect the efficacy of the initial rovibrational excitation. In other words, whereas fast IVR precludes mode selectivity, it facilitates the unraveling of the vibrational molecular dynamics.

Vibrationally Mediated Photodissociation
By Salman (Zamik) Rosenwaks
© Salman (Zamik) Rosenwaks 2009
Published by the Royal Society of Chemistry, www.rsc.org

It seems, indeed, that VMP studies have concentrated recently on molecular dynamics rather than on mode selectivity. We note, however, that the often-observed difference in IVR lifetimes between different modes may also be considered, in a sense, as a reflection of mode selectivity. Working with a better spectral or temporal resolution may demonstrate this selectivity in more detail.

Although our emphasis in this monograph has been on VMP of isolated, neutral molecules in the gas phase, some other applications of VMP are worth noting. For example, VMP of the $V^+(OCO)$ molecular ion is reported in Ref. 1 and of formic acid dimer, $(DCOOH)_2$, in Ref. 2.

A different, but related area of research is on controlling bimolecular reactions *via* vibrational excitation. This area certainly deserves a monograph of its own. We cite here one example[3] in which the reaction of vibrationally excited CH_4 with Cl atoms is discussed and references to earlier studies in this area are cited. Another related area of research is on the influence of vibrational preparation on gas-surface reactivity. An example is the state-resolved gas-surface scattering measurements that reveal that the v_1 C–H stretch vibration in CHD_3 selectively activates C–H bond cleavage on Ni surface.[4]

Our final note is on quantum control of photodissociation. This subject is presented in detail in Refs. 5 and 6. There are two main paradigms for quantum control. (1) Optimal control,[7–11] which is basically a trial-and-error method: One guesses a Hamiltonian where external fields, *e.g.*, laser light, are introduced, does an experiment or propagates the initial wavefunction to the future, compares the result with the desirable objective, and corrects the guess for the Hamiltonian until satisfactory agreement with the objective is reached.[6] (2) Coherent control, where one identifies quantum interferences between simultaneous indistinguishable pathways to the same final state. Optimal control is based on localized wavepackets and apply ultrashort (in the sub-ps and fs timescales), often shaped, laser pulses.[7–11] However, as shown in Ref. 5, it also relies upon the existence of multiple interfering pathways. In both paradigms the application of simultaneous laser pulses (or of one pulse) circumvents IVR that limits mode-selective photodissociation in VMP.

References

1. M. Citir and R. B. Metz, *J. Chem. Phys.*, 2008, **128**, 024307.
2. Y. H. Yoon, M. L. Hause, A. S. Case and F. F. Crim, *J. Chem. Phys.*, 2008, **128**, 084305.
3. H. Kawamata, S. Tauro and K. Liu, *Phys. Chem. Chem. Phys.*, 2008, **10**, 4378.
4. D. R. Killelea, V. L. Campbell, N. S. Shuman and A. L. Utz, *Science*, 2008, **319**, 790.
5. M. Shapiro and P. Brumer, *Principles of the Quantum Control of Molecular Processes*, Wiley, New York, 2003.
6. H. Fielding, M. Shapiro and T. Baumert, ed., *Journal of Physics B: Atomic, Molecular and Optical Physics, Vol 41, Number 7, Special Issue on Coherent Control*, The Institute of Physics, Bristol, 2008.

7. D. J. Tannor and S. A. Rice, *J. Chem. Phys.*, 1985, **83**, 5013.
8. D. J. Tannor, R. Kosloff and S. A. Rice, *J. Chem. Phys.*, 1986, **85**, 5805.
9. S. Shi, A. Woody and H. Rabitz, *J. Chem. Phys.*, 1988, **88**, 6870.
10. W. Jakubetz, J. Manz and V. Mohan, *J. Chem. Phys.*, 1989, **90**, 3686.
11. S. A. Rice and M. Zhao, *Optical Control of Molecular Dynamics*, Wiley, New York, 2000.

Note added in proofs: The following papers, related to VMP theory and practice, have been published in late 2008 and the beginning of 2009 after the completion of this manuscript:

J. A. Beswick and R. N. Zare. "On the quantum and quasiclassical angular distributions of photofragments", *J. Chem. Phys.*, 2008, **129**, 164315. (Relevant to Chapter 2).

W. Lei and C. Jagadish, "Lasers and photodetectors for mid-infrared 2–3 mm applications", *J. Appl. Phys.*, 2008, **104**, 091101. (Relevant to Chapter 3).

A. L. Devine, M. G. D. Nix, R. N. Dixon and M. N. R. Ashfold, "Near-ultraviolet photodissociation of thiophenol," *J. Phys. Chem. A*, 2008, **112**, 9563. (Relevant to Chapter 8).

M. Grechko, P. Maksyutenko, N. F. Zobov, S. V. Shirin, O. L. Polyansky, T. R. Rizzo and O. V. Boyarkin, "Collisionally-assisted spectroscopy of water from 27000 to 34000 cm^{-1}", *J. Phys. Chem. A*, 2008, **112**, 10539. (Relevant to Chapter 6).

W. Lai, S. Y.Lin, D. Xie and H. Guo, "Full-dimensional quantum dynamics of $\tilde{A}$-state photodissociation of ammonia: Absorption spectra", *J. Chem. Phys.*, 2008, **129**, 154311. (Relevant to Chapter 7).

O. P. J. Vieuxmaire, Z. Lan, A. L. Sobolewski and W. Domcke, "*Ab initio* characterization of the conical intersections involved in the photochemistry of phenol", *J. Chem. Phys.*, 2008, **129**, 224307. (Relevant to Chapter 8).

P. E. Crider, L. Castiglioni, K. E. Kautzman and D. M. Neumark, "Photodissociation of the propargyl and propynyl (C_3D_3) radicals at 248 and 193 nm", *J. Chem. Phys.*, 2009, **130**, 044310. (Relevant to Chapter 8).

A. Golan, N. Mayorkas, S. Rosenwaks and I. Bar, "A new method for determining absorption cross-sections out of initially excited vibrational states", *J. Chem. Phys.*, 2009, **130**, 054303. (Relevant to Chapter 8).

K. L. Wells, G. Perriam and V. G. Stavros, "Time-resolved velocity map ion imaging study of NH_3 photodissociation," *J. Chem. Phys.*, 2009, **130**, 074308. (Relevant to Chapter 7 and the Concluding Remarks). This paper points to the importance and possibility of studying sub-ns VMP, alluded to in the Concluding Remarks.

Author Index

Abbouti Temsamani, J., 137, 138
Achiba, Y., 137
Adhikaria, S., 92
Adler-Golden, S. M., 64
Ahlers, G., 138
Ahn, D.-S., 184
Akagi, H., 92, 139
Alekseyev, A. B., 50, 185
Alexander, A. J., 49
Alexander, M. H., 93, 185
Allen, W. D., 140
Allinger, N. L., 27
Ambartzumian, R. V., 40, 49, 139, 183, 188
Amstrup, B., 91
Andresen, P., 41, 71, 90, 91, 93
Anderson, J. G., 64
Aoyagi, M., 67
Arusi-Parpar, T., 41, 91, 136, 137
Ascenzi, D., 49, 185
Ashfold, M. N. R., 27, 41, 42, 65, 90, 137, 138, 139, 140, 183, 185, 186, 187
Astholz, D. C., 64
Atabek, O., 64

Bach, A., 41, 139, 186
Bacis, R., 93
Back, R. A., 139
Badcock, C. C., 50
Baeck, K.-K., 184
Baek, S. J., 183, 184

Baer, M., 28, 184
Baggott, J. E., 27, 28, 42, 90, 140, 186
Bair, E. J., 40, 49, 64, 65
Bairagi, D. B., 50
Balint-Kurti, G. G., 27, 49, 50, 65, 66, 67, 90, 92
Balko, B. A., 137
Ball, S. M., 64
Baloitcha, E., 65
Band, Y. B., 27, 67
Bar, I., 8, 41, 91, 92, 93, 136, 137, 184, 185, 186, 187
Barnes, R. J., 28, 92, 141
Barry, M. D., 185
Bartlett, N. C.-M., 49
Baumert, T., 190
Bayliss, N. S., 49
Baylor, L. C., 187
Bechtel, H. A., 41, 49
Beck, C., 139
Ben-Nun, M., 27
Berghout, H. L., 139, 140
Bergmann, K., 41
Bersohn, R., 27, 66, 93, 140, 185, 187
Bespechansky, E. 187
Beswick, J. A., 50, 67
Bethardy, G. A., 184
Beushausen, V., 90
Biennier, L., 138
Bierbaum, V. M., 187

Biesner, J., 138
Billing, G. D., 66
Bingemann, D., 186
Bischel, W. K., 65, 66, 185
Bittererová, M., 140
Bixon, M., 28
Black, G., 65
Blockstein, L., 187
Bokor, J., 66
Bouchardy, P., 91
Bourgeois, M. J., 64
Bowman, J. M., 92
Boyarkin, O. V., 28, 41, 92
Bramley, M. J., 137
Brickman, R. O., 188
Brion, J., 65
Brom, A. J. van den, 66
Brouard, M., 66, 91, 92, 140
Brown, A., 49, 50, 66, 67, 186
Brown, D. J. A., 41, 49
Brown, S. S., 139, 140
Browning, P. W., 183
Brownsword, R. A., 140, 184
Brumer, P., 8, 27, 190
Buenker, R. J., 50, 185
Burak, I., 67
Burghardt, I., 28
Burkholder, J. B., 40, 49, 64, 65
Burns, G., 49
Butler, L. J., 67, 140, 183, 186

Caldwell, J. J., 137
Callegari, A., 28
Callender, R. H., 49
Camden, J. P., 41, 49
Campargue, A., 137, 138, 187
Campargue, R., 66
Campbell, V. L., 190
Cao, J., 28
Carrasquillo, M., 138
Carter, S., 27, 137
Case, A. S., 186, 190
Cederbaum, L. S., 27, 28
Chandler, D. W., 66
Chang, A. H. H., 187
Chang, Y. C., 41, 67

Cheatum, C. M., 139, 140
Chen, F.-T., 138
Chen, S., 50
Chen, X., 41, 184, 185, 186
Chen, Y. T., 187
Chestakov, D. A., 42, 50
Chien, T. S., 137
Child, M. S., 28, 92, 138
Choi, J.-M., 184
Choi, K.-W., 183, 184
Choi, S., 184
Choi, Y.S., 183, 184
Chou, L.-C., 138
Christoffel, K. M., 92
Citir, M., 190
Clark, R. J. H., 41
Clark, T., 27
Coffey, M. J., 140
Cohen, Y., 41, 91
Cohen-Tannoudji, C., 27
Cox, D. M., 188
Colbert, D. T., 140
Collins, M. A., 27
Comes, F. J., 93, 141
Condon, E. U., 43, 49
Cook, P. A., 185
Cottrill, E. E. H., 187
Crim, F. F., 7, 8, 41, 90, 91, 92, 138,
    139, 140, 186, 190
Croce, A. E., 64
Cronin, B., 186
Cross, P. C., 28
Cui, Q., 138
Cyr, D. R., 40, 50

Dagdigian, P. J., 28, 93, 141, 184,
    186
Dallarosa, J., 66
Darnton, L. A., 65
Daud, M. N., 66
Daumont, D., 65
David, D., 41, 91, 93
Decius, J. C., 28
Delgado, R., 139
Denzer, W., 64, 65
DePuy, C. H., 187

DeSain, J. D., 187
Deshpandea, S., 92
Devine, A. L., 186
Dirke, M. von, 27
Diu, B., 27
Dixon, R. N., 93, 138, 139, 183, 186
Domcke, W., 27, 28, 186
Donovan, R. J., 65
Dorfman, G., 185
Dove, J. E., 66
Dübal, H.-R., 140, 185
Dulligan, M., 41, 49, 136
Dullinger, R. E., 49
Duncan, J. L., 28, 186, 187, 188
Dunn, K. M., 183
Dylewski, S. M., 65
Dymond, E. G., 49

East, A. L. L., 140
Eberly, J. H., 41
Eggleston, J., 185
Ehara, M., 93
Einfeld, T., 41, 185
Elghobashi, N., 92
El Idrissi, M. I., 138
Engel, V., 90, 91
Eppink, A. T. J. B., 50, 185
Evleth, E. M., 183

Fang, W-H., 186
Farantos, S. C., 27, 65
Fasold, R., 141
Felder, P., 184
Feltham, E. J., 187
Field, R. W., 8, 28, 41, 93, 137, 185
Fielding, H., 190
Fitzwater, D. A., 139
Fleischhauer, M., 41
Flöthmann, H., 139
Flynn, G. W., 93
Foo, P. D., 137
Francisco, J. S., 186
Franck, J., 43, 49
Frederick, J. H., 90
Freed, K. F., 27, 67
Fritz, M. D., 138
Fujisaki, H., 50

Fukazawa, H., 92
Fuke, K., 139
Fusina, L., 138
Fusti-Molnar, L., 67

Gabriel, H., 49
Gaillot, A.-C., 138
Gallagher, J. Y., 64
Gallego, A. P., 187
Ganot, Y., 41, 137, 186, 187
Garcia, I. A., 40, 42, 50
García, R. P., 187
Gaspard, P., 137
Gasteiger, J., 27
Gericke, K.-H., 41, 66, 93, 141, 185
Gersonde, I. H., 49
Gersten, J. I., 49
Gibson, G. E., 49
Gilles, M. K., 65
Givertz, S. C., 50
Gleghorn, J. T., 183
Golan, A., 41, 137, 184, 186
Goldstein, H., 27
Goncher, S. J., 187
Gordon, R. J., 185
Gorman, M. P., 186
Gorry, P. A., 185
Grant, E. R., 93
Grebenshchikov, S. Yu., 65
Green, A. V., 66
Green, C. H., 28
Greenslade, M. E., 50
Grisch, F., 91
Groenenboom, G. C., 40, 50, 66
Gross, A. F., 92, 141
Gross, P., 50
Gruebele, M., 28
Gruodis, A., 66
Guan, Y., 140
Guo, H., 67, 185
Gupta, A. K., 50

Halász, G. J., 184
Halfmann, T., 41
Hall, G. E., 66, 185
Halonen, L., 28, 92, 138

Halpern, J. B., 138
Hamilton, C. H., 41
Hammerich, A. D., 185
Hancock, G., 64, 65
Handy, N. C., 137
Hanson, H. G., 49
Harrevelt, R. van, 92
Hartke, B., 90
Hashimoto, N., 138
Hashimoto, S., 185
Hause, M. L., 139, 186, 190
Häusler, D., 41, 90
Heck, A. J. R., 66
Heflinger, D., 91
Heller, E. J., 27
Hemert, M. C. van, 92
Henning, S., 49, 90, 91
Henriksen, N., 91
Henry, B. R., 28
Herman, M., 137, 138, 187
Herschbach, D. R., 49
Herzberg, G., 8, 28, 40, 67, 92, 138,
    139, 140, 183
Hessel, M. M., 49
Hester, R. E., 41
Heyst, J. van, 66
Hidaka, Y., 187
Higashihara, T., 187
Hillenkamp, M., 140, 184
Hioe, F. T., 41
Hippler, H., 66
Hippler, M., 186
Hirst, D. M., 27
Hofmann, P., 187
Holiday, R. J., 41, 139, 186
Holland, S. M., 140
Horsley, J. A., 188
Horrocks, S. J., 65
Houston, P. L., 65, 67, 185
Howard, B. J., 93
Hsiao, M. C., 91
Hsieh, S., 140, 186
Hsu, Y.-C., 138
Hsu, Y.-T., 138, 184
Huang, J. D., 187
Huber, J. R., 27, 93, 184

Hudson, B., 91
Hutchison, J. M., 41, 139,
    186
Huxley, P., 27
Hwang, D. W., 187
Hwang, W. C., 50
Hynes, J. T., 28

Imre, D. G., 90, 92
Ingold, C. K., 137
Innes, K. K., 137
Iogansen, A. A., 188
Iwata, S., 137
Iwamoto, S., 137

Jackson, W. M., 138, 187
Jacon, M., 64
Jakubetz, W., 191
Janssen, L. M. C., 40, 50
Janssen, M. H. M., 41, 66
Jezowski, S. R., 66
Joens, J. A., 49, 64, 65
Johnson, C. S., 140
Johnson, H. S., 66
Johnson, M. S., 66
Johnston, G. W., 187
Jortner, J., 28, 91
Judge, D.L., 137
Jung, K.-H., 184
Jung, Y. J., 184

Kaldor, A., 188
Kalyanaraman, C., 50
Kamada, R. F., 50
Kash, P. W., 67, 186
Kassab, E., 183
Katayama, D. H., 64
Katayanagi, H., 66, 67
Kawamata, H., 66, 190
Kawano, H., 187
Kawasaki, M., 64, 65, 66, 185
Kaye, J., 184
Kaya, K., 139
Kerridge, B. J., 50
Killelea, D. R., 190
Kim, M. H., 66
Kim, S. K., 183, 184

Kimura, K., 137
King, A. M., 186
King, D. S., 93
King, G. W., 137
Kinsey, J. L., 41, 93
Kishigami, M., 64
Kitchen, D. C., 183
Kitsopoulos, T. N., 42, 66
Kleindienst, T. E., 64
Kleinermanns, K., 93
Klemperer, W., 138
Kligler, D. J., 65
Klossika, J.-J., 139, 140
Knight, A. E., 92
Knupfer, P., 184
Koda, S., 139
Kohguchi, H., 66, 138
Kokh, D. B., 50
Kollman, P. A., 27
Komissarov, A. V., 66
Kono, M., 183
Köppel, H., 27, 28
Kosloff, R., 184, 191
Kotov, A. V., 188
Koubenakis, A., 50
Kraemer, G. T., 50
Krause, P., 92
Kreglewski, M., 184
Kuchitsu, K., 93
Kung, A. H., 187
Kurkala, V., 92

Lai, L.-H., 184
Laloe, F., 27
Lambert, H. M., 28, 141, 184, 186
Lamotte, J., 187
Landau, L. D., 8
Lan, Z., 186
Lang, V. I., 65
Langford, S. R., 91, 92, 185
Lao, K. Q., 185
Laurent, T., 140, 184
Läuter, A., 92
Lavally, J. C., 187
Law, D. W., 28
Law, M. M., 28, 186

Lawley, K. P., 65
Lawton, R. T., 92
Le Roy, R. J., 41, 49, 50
Leahy, D. J., 40, 50
Lee, J., 184
Lee, K., 184
Lee, K.-S., 184
Lee, S., 50
Lee, S. K., 66
Lee, Y. S., 183
Lee, Y. T., 42, 137, 187
Lehmann, K. K., 28, 138
Leigh, R. W., 49
Leitner, D. M., 28
Lester, M. I., 50
Letokhov, V. S., 40, 49, 139, 183
Leung, K. T., 185
Levi, C., 184, 187
Levine, R. D., 8, 27, 91
Li, L., 185
Li, R.-J., 136, 137
Li, W., 66
Liebermann, H. P., 50, 185
Lievin, J., 137, 187
Lifshitz, E. M., 8
Lightfoot, P. D., 28
Likar, M. D., 140, 186
Lim, J. S., 183
Lineberger, W. C., 187
Liou, H. T., 41, 67
Liou, Z., 41, 67
Lipciuc, M. L., 41, 66
Little, D. D., 41, 49, 64, 65, 186
Liu, K., 138, 184, 190
Liu, G. S., 137
Lin, S. H., 187
Liu, Y-J., 186
Liyange, R., 185
Lock, M., 141
Locker, J. R., 40, 65
Loffler, P., 138
Lonardo, G. D., 138
Longfellow, C. A., 65
Loo, M. P. J. van der, 40, 50
Loock, H-P., 28
Luckhaus, D., 186

Lülf, H. W., 90
Luntz, A. C., 93

Mabbs, R., 140
Macdonald, R. G., 49
Mahapatra, S., 186
Maier, R. J., 65
Makarov, G. N., 40, 139
Maksyutenko, P., 28, 41, 92
Malicet, J., 65
Mallinson, P. D., 188
Manolopoulos, D. E., 91
Manz, J., 90, 92, 191
Marom, R., 41, 184, 185, 186
Marston, C. C., 186
Martin, J. M. L., 184
Martin, M. R., 49
Martin, S. E., 64
Martínez, T. J., 27
Martinez, M. T., 140
Martinez-Avilés, M., 186
Martinez-Nunez, E., 187
Mashino, M., 66
Mastenbroek, J. W. G., 41
Masturzo, D. E., 41, 65
Mathis, J., 90
Matsumi, Y., 64, 65, 66, 185
Matthews, J., 186
Maul, C., 41, 185
McCaffery, A. J., 93
McDade, I. C., 64
McDonald, J. M., 67
McFarlan, M., 184
McGlynn, S. P., 67
McGrath, W. D., 64
McKean, D. C., 186, 188
McLaren, I. H., 41
Mebel, A. M., 187
Meier, W., 41
Meiert, U., 141
Melchior, A., 41, 184, 185
Metz, R. B., 139, 190
Michael, J. V., 183
Michelsen, H. A., 64
Mills, I. M., 28, 137, 140
Minitti, M. P., 66

Mishra, M. K., 50, 92
Mitsuhashi, F., 137
Mittal, J. P., 92
Mohan, V., 191
Monnerville, M., 50
Morokuma, K., 138, 183
Morse, M. D., 67
Mourdaunt, D. H., 137, 138, 139, 183
Muenter, J. S., 41, 92
Mulliken, R. S., 185
Murrell, J. N., 27, 92
Myers, T. L., 92

Nahler, N. H., 42, 187
Naik, P. D., 92
Nakajima, A., 139
Nakamura, H., 50
Nan, G., 67
Nanbu, S., 66, 67
Nesbitt, D. J., 8, 28, 41, 91, 92, 137
Neumark, D. M., 40, 50, 187
Newnham, D. A., 140, 186
Neyer, D. W., 66
Ni, C.-K., 187
Nishide, T., 66
Nix, M. G. D., 186
Nizkorodov, S. A., 92
Noyes, W.A., 183

Obi, K., 185
Offer, A. R., 67
Okabe, H., 67, 185, 187
O'Keeffe, P., 65
Oki, T., 187
O'Mahony, J., 140
Opel, M., 92
Oref, I., 187
Oreg, J., 41
Orr, B. J., 136
Orr-Ewing, A. J., 42, 49, 50, 64, 185
Osamura, Y., 137
Osborn, D. L., 40, 50

Palma, A., 92
Papernov, S. M., 39

Park, M. H., 184
Parker, D. H., 40, 42, 50, 185
Pate, B. H., 28
Paur, R. J., 64
Pealat, M., 91
Pearson, P. J., 65
Perry, D. S., 28, 184
Person, M. D., 186
Pfab, J., 41
Pinot de Moira, J. C., 64, 65
Pisarchik, A., 137, 138
Plach, H. J., 66
Plusquellic, D. F., 91, 92
Polak, M. L., 187
Polyansky, O. L., 41, 92
Pomerantz, A. E., 41, 49
Portnov, A., 41, 137, 184, 186, 187
Pouilly, B., 50, 93
Powis, I., 28
Pullman, B., 91
Pummer, H., 65, 185
Puretzkiy, A. A., 40, 139
Pyle, J. A., 50

Qadiri, R.H, 187
Qian, C. X. W., 28
Qu, Z.-W., 65
Quack, M., 28, 185, 186
Quadrini, F., 66

Rabalais, J. W., 67
Rabinowitz, P., 188
Rabitz, H., 191
Radenović, D. Č., 40, 50
Rakitzis, T. P., 49, 50, 66, 138
Ravishankara, A. R., 65
Reed, C.L., 183
Regan, P. M., 49, 185
Reid, C., 140
Reimers, J. R., 92
Reisler, H., 93, 140
Rhodes, C. K., 65, 66, 185
Rice, O. K., 49
Rice, S. A., 191
Ridley, T., 65
Riehn, C. W., 41, 49, 136

Riley, S. J., 42
Rist, C., 185
Ritchie, G. A. D., 65
Rizzo, T. R., 28, 41, 92
Robiette, A. G., 28
Robinson, J. C., 187
Robinson, M. S., 187
Roij, A. J. A. van, 40, 50
Rosenwaks, S., 8, 41, 91, 92, 93, 136,
      137, 184, 185, 186, 187
Rothe, E. W., 93
Rowland, F. S., 184
Rubio-Lago, L., 49, 50

Sadowski, C. M., 64
Salawitch, R. J., 64
Sarma, M., 92
Sathyamurthy, N., 50
Satyapal, S., 93, 187
Saussey, J., 187
Schaefer III, H. F., 27
Schatz, G. C., 67, 185
Scherer, G. J., 138
Scherr, V., 67
Schinke, R., 7, 27, 41, 50, 65, 90, 91,
      93, 139, 140, 186
Schleyer, P. v. R., 27
Schmeer, J., 138
Schmid, R.P., 136, 137
Schmiechen, P., 184
Schneider, L., 41, 138
Schreiner, P. R., 27
Schröder, T., 93
Schweitzer, E. L., 64
Scoles, G., 28
Scott, J. L, 41, 90, 91, 186
Segev, E., 90
Seki, K., 187
Selwyn, G. S., 66
Sension, R. J., 91
Settle, R. D. F., 28
Shafer, N., 93
Shapiro, M., 8, 27, 50, 66, 67, 90, 91,
      92, 93, 185, 190
Sharma, R., 186
Sharon, R., 67

Sharpless, R. L., 65
Sheng, X., 41, 137
Sheppard, M. G., 64
Shi, S., 191
Shi, Y., 138
Shimanouchi, T., 67, 92, 137, 184
Shirin, S. V., 41, 92
Shiromaru, H., 137
Shiu, Y.-J., 138
Shlyapnikov, G. V., 49
Shore, B. W., 41
Shuman, N. S., 190
Sibert III, E. L., 140
Simons, J. P., 93, 140
Simons, J. W., 64
Sinha, A., 28, 91, 92, 140, 141, 186
Smith, E. W., 49
Sobolewski, A. L., 186
Sofikitis, D., 49, 50
Sorbie, K. S., 92
Speirs, G. K., 187
Staemmler, V., 90, 91, 92, 141
Stanton, J. F., 138
Stein, A., 188
Steinfeld, J. I., 64
Sticklandt, R. J., 140
Stolte, R., 41
Strugano, A., 41, 91, 93
Sugita, A., 66
Suits, A. G., 66, 187
Sun, W., 187
Suter, H. U., 27, 186
Suzuki, T., 66, 67, 138
Sveum, N. E., 187

Taatjes, C.A., 187
Takahashi, K., 64, 65
Talukdar, R. K., 65
Taniguchi, N., 64, 65, 187
Tannor, D. J., 191
Tao, C., 186
Taran, J. P., 91
Tasaki, S., 185 K.
Tashiro, M., 65
Tauro, S., 190
Teller, E., 67

Ter Meulen, J. J., 50
Teranishi, Y., 50
Teule, J. M., 66
Thanopulos, I., 185
Theuer, H., 41
Thompson, D. L., 140
Ticich, T. M., 140, 186
Tobiason, J. D., 138
Tonokura, K., 185
Toomes, R. L., 42
Toumi, R., 50
Troe, J., 64, 66
Trott- Kriegeskorte, G., 66
Truhlar, D. G., 27
Tsuchiya, S., 137
Tsuji, K., 185
Tsukamoto, K., 139
Tyley, P., 64, 65

Underwood, J. G., 28
Untch, A., 27
Upadhyaya, Á. H. P., 184
Utz, A. L., 138, 190

Valentini, J. J., 41, 91
Vallance, C., 66
Vallet, V., 186
Vandana, K., 50
Vander Auwera, J., 137, 187
Vander Wal, R. L., 41, 90, 91, 140, 186
Varandas, A. J. C., 27
Vasudev, R., 93
Vasyutinskii, O. S., 50
Vatsa, R. K., 140, 184
Vazquez, S. A., 187
Vibók, Á., 184
Vieuxmaire, O. P. J., 42
Vitanov, N. V., 41
Volpp, H.-R., 92, 140, 184
Votava, O., 91, 92

Wada, A., 139
Walker, R. B, 64
Wang, J.-H., 138
Wang, X .L., 184

Waschewsky, G. C. G., 183
Watts, J. R., 92
Webster III, H . A., 40, 64, 65
Weide, K., 41, 90, 91, 186
Weitz, E., 187
Welge, K. H., 41, 138
Wennberg, P. O., 64
Western, C., 41
Wiesenfeld, J., 93
Wiley, W. C., 41
Wilson, E. B., 28
Wilson, K. R., 42
Wilson, S.H. S., 138
Winterbottom, F., 64
Wittig, C., 41, 49, 93, 136
Wodtke, A. M., 42, 137
Wolfrum, J., 92, 140, 184
Wolynes, P. G., 28
Woods, E., 140
Woody, A., 191
Wrede, E., 138
Wu, C. Y. R., 137
Wu, S-M., 40, 42, 50

Xiao, H-Y., 186
Xie, X. , 138

Yamada, K., 140
Yamakita, N., 137
Yamashita, K., 93, 139
Yang, J. L., 49

Yang, X., 184
Yang, X. M., 187
Yang, Y., 185
Yanson, M. L., 49
Yarkony, D. R., 28, 139
Yi, W. K., 140
Ying, J. F., 185
Yokoyama, A., 92, 139
Yokoyama, K., 92, 139, 187
Yonekura, N., 138
Yoo, H.-S., 184
Yoon, Y. H., 139, 186, 190
Yoshida, Y., 139
Yu, J-G., 186
Yuan, Y., 50
Yurchenko, S. N., 185

Zare, R. N., 28, 41, 49, 93, 185
Zecharia, U., 184
Zeiri, Y., 184
Zhang, J., 41, 49, 50, 90, 92, 136, 137
Zhao, M., 191
Zhou, W., 50
Zhu, H., 65
Ziemkiewicz, M., 92
Zittel, P. F., 41, 49, 64, 65, 186
Zobov, N. F., 41, 92
Zubarev, I. G., 188
Zwielly, A., 187
Zyrianov, M., 140

Subject Index

absorption cross sections 15, 16, 45, 52, 71, 72
acetylene
  homologues 172–177
  isotopologues 94–110
action spectrum 5, 29–31, 36, 37, 82, 97, 99–104, 106–110, 112–114, 117, 122, 123, 145–148, 150, 152, 166, 170, 172, 173, 176, 178–180
adiabatic
  electronic states 11, 19
  PES 11, 12, 20, 96, 143
  representation 11, 12
  VMP 48, 74, 96–104, 107–110, 115, 116, 127, 150, 166
ammonia isotopologues 110–120
angular distributions 10, 20, 39, 43, 44, 47, 88, 159, 161
anharmonicity 22–24
anisotropy parameter 20, 21, 54, 59, 136, 155, 156, 159, 161, 164
atmospheric chemistry 6, 53, 54, 152, 171, 172

bond-selective dissociation 5–7, 43, 51, 129, 142
Born-Oppenheimer (BO)
  approximation 10–12, 15, 20
tert-butyl hydroperoxide 169
1-butyne 174–177

CD_3NH_2 150, 151
CF_3I 163

CHD_2Cl 163, 164
$CHFCl_2$ 152–156
CH_2Cl_2 164
CH_3CFCl_2 156, 157
CH_3Cl 163, 164
CH_3I 157–162
CH_3NH_2 143–150
CH_3OH 26, 27, 169, 170
CH_3OOH 172
$(CH_3)_3COOH$ 169
CS_2 61, 62
C_2HD 107–110
C_2H_2 94–107
C_2H_4 177–180
Chappuis absorption band 52, 53
Cl atoms 45, 152, 153, 155
classical trajectory approach 14
coherent anti-Stokes Raman spectroscopy (CARS) 34-36, 82, 83
coherent control 162, 190
cold molecular beams 58, 59
conical intersections (CI) 18–20, 110–116, 144, 145, 149, 151, 159, 163, 165, 166
Coriolis coupling 24, 104, 178

D atoms 47, 69, 118, 151, 164, 173
DCN 63, 64
DOCl 63, 64
D_2O 90
$D_3CC{\equiv}CH$ 173, 174
Darling–Dennison resonance 24, 152
diabatic 11, 12, 48, 52, 143, 161, 162

diatomic molecules 43–49
diatomic radicals 47, 48
direct photodissociation 3, 4
dissociation energy of water 81
Doppler profiles 30, 38, 115, 136,
    148–151
double-resonance 80, 81

eigenstate 16, 23, 24,26
    molecular 23, 24, 70
    vibrational 23, 82, 163
    zero-order 23, 24
electronic excitation 36, 37
ethene isotopologues 177–182
excitation
    electronic 36, 37
    vibrational 32–36

Fermi
    Golden Rule 14, 25
    resonance 24, 150, 152, 176, 177,
        179, 180
far off-resonance excitation 74, 75
"Feshbach-type" rotational resonances 70
Franck-Condon (FC) 74, 77–79, 83,
    86
    factors 4, 53, 54, 56, 71, 73, 83, 84,
        101, 103, 105–107, 110, 112, 117,
        122, 123, 129, 147, 148, 153, 156,
        176
    mapping 73
    models 15, 71, 76, 77, 86, 172
    overlap 3, 5, 15, 47, 55, 72, 84, 88,
        103, 106, 110, 131, 133, 147, 156,
        161, 179, 180
    point 11, 14, 17
    principle 3, 4, 6, 10, 14, 43
    pumping 32, 35
    region 72, 74, 87, 89, 97, 143, 162,
        164, 170
    rotational factors 15
    vertical regime 136

H atoms 44, 46, 47, 69, 97, 103, 106,
    115, 118, 144, 145, 148–152,
    164–166, 173–175

HBr 45
HCN 63, 64
HCl 44, 45
trans-HDC=CDH 178, 179, 180–182
HF 44
HI 46, 47
HNCO 12, 13, 120–129
HN_3 135, 136
HOBr 63, 64
HOCl 63, 64
HOD 81–89
HONO 134, 135
$HONO_2$ 168
HOOD 131–134
HOOH 131–134
HO_2NO_2 171, 172
$H_2C=CD_2$ 178, 179
$H_2C=CH_2$ 177–182
H_2O 33, 68–81
$H_3CC \equiv CH$ 174, 176, 177
$H_3CH_2CC \equiv CH$ 174–177
haloalkanes 151–164
halogen atoms 151
Hamiltonian 11, 16, 22, 23, 24, 26, 70,
    177, 181, 182, 190
    electronic 11, 12
    LM/NM 150
    molecular 11, 23, 25
    vibrational 23, 82, 163
    zero-order 23, 24
Hartley absorption band 52, 53
Herzberg-Teller coupling 55
hexapole state selection 33, 47, 58,
    59, 61
high-n Rydberg time-of-flight (HRTOF)
    38, 39, 44, 47, 97, 144, 174
Huggins absorption band 52, 53
hydrazoic acid 135, 136
hydrochlorofluorocarbons 151–157
hydrogen peroxide isotopologues
    131–134
hydroxylamine 170, 171

I atoms 47, 158, 159, 161
ICN 63
indirect photodissociation 3, 4

internal conversion (IC), 19, 32, 96, 101, 128, 129, 145, 150, 151, 165, 166, 173–178

intersystem crossing, 19, 96, 101, 120, 122, 128

intramolecular vibrational redistribution (IVR) 3, 10, 21–27, 93, 123, 134, 142, 144, 147, 148, 173, 176, 180, 188, 189

isocyanic acid 12, 13, 120–129

isotope separation 5, 58, 81, 96, 107, 119, 150, 151, 179, 183

Λ-doublet distributions 75, 76, 78, 79

laser-induced fluorescence (LIF) 2, 5, 21, 29, 37, 53, 69, 70, 71, 79, 80–85, 88, 90, 122–131, 133, 135, 168, 170, 172

lifetime 10, 20, 25, 26, 88, 97, 117, 118, 126, 127, 144, 148, 151, 152, 156, 189, 190

local modes (LM) 22, 23, 70–72, 74, 76, 80, 88, 131, 133, 150, 164, 176, 177, 179

methanol 26, 27, 169, 170

methyl hydroperoxide 172

methyl iodide 157–162

methylamine isotopologues 143–151

mode-selective dissociation 5–7, 43, 51, 129, 142, 189

Morse oscillators 22, 23

NHD$_2$ 118–120

NH$_2$ stretch state 149, 150

NH$_2$D 118–120

NH$_2$OH 170, 171

NH$_3$ 111–118

N$_2$O 60, 61

NaI 43

nitric acid 168

nitrous acid 134, 135

nonadiabatic
  electronic states 53, 96
  transitions 12, 19, 20, 161
  VMP 96, 104–107, 110, 114–116, 120, 127, 149, 163, 164

normal modes (NM) 22, 23, 61, 70, 95, 96, 120, 134, 143, 150, 159, 160, 164, 176, 179–181

O atoms 47, 53, 60

OCS 55–60

OCSe 60

OD radicals 47

OH radicals 47, 73–79

O$_2$ 48, 49

O$_3$ 51–55

on-resonance excitation 74, 75

one-photon excitation 32, 33

optimal control 190

ozone 51–55

peroxynitric acid 171, 172

phenol 164–166

photoacoustic (PA) spectroscopy 31, 35, 36, 105, 112, 113, 117, 123, 150, 152, 164, 172, 176-180

photoacoustic Raman spectroscopy (PARS) 34-36, 122, 145-147

photodissociation 3–5
  cross sections 13–18, 82, 83
  dynamics 9–21

photofragment translational spectroscopy (PTS) 39, 144, 174

photofragments, detection 37–40

polyads 70–75, 152–154

potential-energy surfaces (PES) 4–6, 10–21, 43, 69, 74, 78–81, 86, 88, 89, 90, 104, 115–120, 131, 134, 143, 147, 173, 174, 176, 178, 179

propyne 174, 176, 177

propyne-d_3 173–174

quantum control 190

radiationless transitions 18, 129

radicals 47–48

resonantly enhanced multiphoton ionization (REMPI) 29-31, 37, 38, 145-147, 152, 153, 165, 166, 170, 173, 177

rotational distributions 58, 63, 64, 70, 71, 75–79, 86, 98, 100–102, 107–110, 133, 135

S atoms 47, 48, 56–58

SD radicals 47, 48

SH radicals 47, 48

simulated emission pumping 35

spin-orbit distributions 75, 76, 78–79
state-selected VMP 1–4
state-to-state photodissociation 7, 43, 54,
    59, 60, 68, 71, 76, 77, 79, 81, 94, 112
stimulated emission pumping (SEP) 36, 62
stimulated Raman adiabatic passage
    (STIRAP) 35, 36
stimulated Raman excitation (SRE) 2,
    29, 30, 34, 35, 69, 77, 79, 82, 83,
    122, 128-130, 145-148, 152, 156
supersonic expansion 4
tetratomic molecules 94–136
thermal vibrational excitation 32
thiophenol 142
tier models 25
time-dependent approach 15–18
time-independent approach 14, 15, 18
time-of-flight mass spectrometer (TOFMS)
    31, 37, 38, 44-46. 62, 97, 153, 154
trajectory approach 14
triatomic molecules 51–64, 68–90
triple-resonance 81
two-photon excitation 33–36

uranium hexafluoride (UF_6) 183

vector correlations 20, 21, 54, 55, 59,
    60, 79, 133, 135, 170
velocity distributions 30, 38, 88, 148,
    154, 155, 165
velocity-map imaging (VMI) 30, 39,
    40, 47, 58, 59, 114, 116, 160, 165
vibrational distributions 21, 23, 24,
    26, 32, 63, 73–75
vibrational excitation 32–36

water isotopologues 33, 68–90
wavepacket 16-18, 72, 75, 84, 88,
    116, 117, 120, 164, 169, 190
Wulf absorption band 52–53

zero-order bright states (ZOBS) 23–26,
    180–182
zero-order dark states (ZODS) 23–26,
    181, 182